LUNAR MODULE
LM 10 THROUGH LM 14

VEHICLE FAMILIARIZATION MANUAL

LM PUBLICATIONS SECTION / PRODUCT SUPPORT DEPARTMENT / GRUMMAN AEROSPACE CORPORATION / BETHPAGE / NEW YORK

©2012 Periscope Film LLC
All Rights Reserved
ISBN #978-1-937684-63-1
www.PeriscopeFilm.com

LMA790-2

LUNAR MODULE
LM 10 THROUGH LM 14

THIS PUBLICATION SUPERSEDES LMA790-2 DATED 28 AUGUST 1969

VEHICLE
FAMILIARIZATION MANUAL

PUBLICATIONS SECTION/PRODUCT SUPPORT DEPARTMENT/GRUMMAN AEROSPACE CORPORATION/BETHPAGE/NEW YORK

1 NOVEMBER 1969

LIST OF EFFECTIVE PAGES

INSERT LATEST CHANGED PAGES. DESTROY SUPERSEDED PAGES.

NOTE: The portion of the text affected by the changes is indicated by a vertical line in the outer margins of the page.

Total number of pages in this publication is 154 consisting of the following:

Page No.	Issue
Title Page	Original
A Page	Original
i thru x	Original
1-1 thru 1-9	Original
1-10 Blank	Original
2-1 thru 2-10	Original
3-1 thru 3-117	Original
3-118 Blank	Original
4-1 thru 4-4	Original

*The asterisk indicates pages changed, added, or deleted by the current change.

TABLE OF CONTENTS

LIST OF ILLUSTRATIONS

1 November 1969

Figure	Title	Page

LIST OF TABLES

Table	Title	Page

INTRODUCTION

This Lunar Module Vehicle Familiarization Manual has been prepared as an aid for orientation and indoctrination purposes only. It describes the LM mission, structure, subsystems, and ground support equipment, including modifications being incorporated into LM's 10 through 14 to support increased lunar mission requirements. Coverage applicable only to the modified vehicle is identified in blue. The following table summarizes subsystem changes related to the modification program.

LM Subsystem Change Summary

Subsystem	Modification	Remarks
Structure	S-band erectable antenna moved from quadrant 1 to quadrant 4.	Provides clearance for payload in quadrant 1.
	Relocated ED relay box and pyro battery.	Provides clearance for addition of water and GOX tanks in quadrant 4.
	Modified MESA support structure and deployment mechanism in quadrant 4.	Required for increased size of MESA, and to provide clearance for addition of water and GOX tanks in quadrant 4.
	Batteries and electrical control assemblies relocated to the -Z end of bulkhead.	Provides clearance for payload in quadrant 1 and addition of water and GOX tanks in quadrant 4.
	Modified RCS plume deflectors.	Provide protection for modified MESA.
	Modified support structure of ascent stage.	Required for increase in crew provisions.
	Cleared quadrant 1 for payload.	Equipment required for increasing lunar mission requirements.
	Insulated docking tunnel.	Reduce cabin heat leakage.
Main Propulsion Subsystem	Added controlled bleed vent for descent section supercritical helium tank.	Controlled bleed vent extends maximum standby time of supercritical helium from 131 to 190 hours.
	Possibility of including 10-inch fixed nozzle extension for descent engine.	Additional exit area ratio will provide higher engine efficiency at full throttle firing. (I_{sp} increased by 2.5 seconds.)
	Removed descent propellant tank balance lines.	Reduces inert weight and residual propellants associated with large cg offsets.

LM Subsystem Change Summary (cont)

Subsystem	Modification	Remarks
Reaction Control Subsystem	Removed thruster isolation valves.	Change will afford weight saving of approximately 25 pounds.
Instrumentation Subsystem	Added 10 transducers.	Satisfy additional measurement requirements, as for additional oxygen and water tanks.
	Added three signal-conditioning plug-in sub-assemblies.	Satisfy additional measurement requirements.
Electrical Power Subsystem	Added one descent battery to -Z-end bulkhead.	Changes required to support 54-hour lunar stay.
	Relocated three batteries and two electrical control assemblies on -Z-end bulkhead.	
Environmental Control Subsystem	Removed battery cold rails from quadrants 1 and 4; added shorter cold rails to -Z-end bulkhead.	Descent batteries relocated.
	Added water and GOX tanks in quadrant 4.	Increases capacity for extended lunar stay.
	Added water and GOX check valves.	Isolate individual tanks to minimize possibility of total loss of expendables.
	Separable water, oxygen, and electrical umbilical assemblies.	Enable shirtsleeve mode in cabin.

1 November 1969

LM Subsystem Change Summary (cont)

Subsystem	Modification	Remarks
Environmental Control Subsystem (cont)	Added capability for PLSS oxygen recharge for 1,382 psi.	Supports augmented PLSS configuration.
Crew Provisions	Removed MESA from quadrant 4 in descent stage. Modified MESA for increased capability.	New MESA, externally mounted at quadrant 4, includes storage of all required PLSS expendables, LiOH canisters, crew supplies, and tools.
	Increased modularized crew stowage provisions in ascent stage, as required, for 54 hour lunar stay.	See figure 3-12.1.
	Strengthened structural members.	Support heavier SLSS and PLSS.
	Improved urine and PLSS waste management as follows:	
	- Waste storage container in descent stage	Vented to lunar atmosphere.
	- Urine receptacle in ascent stage	
	- Quick-disconnects for waste transfer	
	- Stowable fecal receptacle	
	Modifications for improved habitability.	See figure 3-12.1.
	Third specimen return container.	In MESA on descent storage; transferred, and stowed behind ascent engine cover during ascent.

LM Subsystem Change Summary (cont)

Subsystem	Modification	Remarks
Displays and Control		
Panel 11	Added URINE LINE circuit breaker.	
	Relocated PROPUL-DES He REG/VENT circuit breaker.	
Panel 2	Modified O$_2$/H$_2$O QTY MON selector switch.	Enables monitoring added second oxygen and water tank in descent stage.
Panel 3	Modified TEMP MONITOR selector switch.	Enables monitoring of MESA temperatures.
Panel 8	Added two Heater Control switches - HTR CONT: MESA and URINE LINE.	Provides temperature stabilization control of MESA and urine line during lunar stay.
Panel 14	Add control switches for fifth descent battery.	Provides controls for lunar stay battery.
	Modified POWER/TEMP MON selector switch.	Enables monitoring of lunar stay battery.
Panel 16	Added MESA circuit breakers.	

1 November 1969

SECTION I

PRELIMINARY MISSION DESCRIPTION

1-1. GENERAL.

A typical mission of the Lunar Module (LM) begins with its separation from the orbiting Command/Service Module (CSM), continues through lunar descent, lunar stay, and lunar ascent, and ends at rendezvous and docking with the orbiting CSM before the return to earth. The LM mission is part of the overall Apollo mission, the objective of which is land two astronauts and scientific equipment on the moon and return them safely to earth.

1-2. EARTH VICINITY AND TRANSLUNAR COAST. (See figure 1-1, sheet 1.)

The Saturn launch vehicle inserts the spacecraft, which is attached to the spacecraft-Lunar Module adapter (SLA), into earth orbit. The LM landing gear is folded and the antennas are stowed while the LM is inside the SLA.

When earth orbit is achieved, the S-IVB stage is shut down and the three astronauts in the Command Module (CM) perform systems status checks and a CSM guidance system reference alignment. Upon completion of earth orbit, the S-IVB engine is restarted to begin translunar injection.

After the initial translunar coasting period, the CSM detaches from the SLA and S-IVB stage, pitches 180°, and docks with the LM - a maneuver called transposition and docking. During this maneuver, the LM/S-IVB stage is stabilized by the S-IVB instrumentation unit. After the CSM pulls the LM free, the S-IVB and the SLA are jettisoned and the spacecraft is oriented for continuation of the translunar coast period. During translunar coast, the LM remains passive, except for the inertial measurement unit (IMU) heaters and portions of the Environmental Control Subsystem (ECS) and Electrical Power Subsystem (EPS), which were activated before launch. The CM performs all navigation and guidance functions and, oriented by the Service Module (SM) reaction controls, initiates midcourse correction maneuvers.

1-3. LUNAR VICINITY. (See figure 1-1, sheet 2.)

Approximately 64 hours after launch, the CSM service propulsion system inserts the spacecraft into an elliptical lunar orbit of approximately 60 by 170 nautical miles. While in this orbit, the astronauts perform CSM guidance system reference alignments and orient the spacecraft attitude for a circularization burn at the beginning of the third lunar orbit. At completion of this maneuver, the spacecraft is in a circular orbit 60 nautical miles above the moon.

The spacecraft is prepared for a docked CSM/LM descent orbit insertion maneuver (DOI) from the 60 nautical-mile circular orbit. There is no LM activity connected with the descent orbit insertion maneuver.

The descent orbit insertion maneuver, which, when completed, will have placed the spacecraft into an elliptical orbit of approximately 9 x 60 nautical miles, is initiated. One-half revolution after initiation of the DOI, the astronauts transfer to the LM to perform activation and checkout procedures. These procedures, which take approximately 4 hours to perform include subsystem activations, IMU alignments, landing gear deployment, and caution/warning checkout. Approximately 2-1/2 revolutions after DOI, the LM undocks and separates from the CSM. Three revolutions after DOI, at approximately 50,000 feet, the CSM returns to the 60 nautical-mile circular orbit; at the same time, the LM begins its descent to the lunar surface.

Descent to the lunar surface consists of three distinct phases: the braking phase from approximately 50,000 to 10,000 feet (high gate), a final approach phase from approximately 10,000 feet to 700 feet (low gate) during which the landing site is observable, and the landing phase, which terminates at touchdown. Descent is performed automatically under control of the Guidance, Navigation, and Control Subsystem (GN&CS) to approximately 700 feet above the lunar surface.

Approximately 2 minutes before reaching the low-gate point, the LM is oriented to begin the final approach phase. During the final approach phase, the LM descends to the low-gate point at nearly constant flight path angle; the attitude is such that the astronauts can observe gross landing area details and manually guide the LM to an alternative landing site, if necessary.

At the low-gate point, the astronauts can select the best landing site and perform the landing phase to touchdown. To accomplish translation to a desired spot on the lunar surface, the thrust vector can be tilted to accelerate the LM in the direction of the landing site. At approximately 3 feet above the lunar surface, the engine is cut off and the vehicle free-falls to the lunar surface.

After touchdown on the lunar surface, the two astronauts perform a lunar surface IMU alignment and check all subsystems to determine whether damage occurred upon landing and to assure that all systems can perform the functions required for a successful ascent. The decision is then made whether the nominal planned stay-time operations can be executed. If all the systems check out satisfactorily, the astronauts observe the surrounding lunar landscape, check the LM hatches, and perform a final check of the portable life support system (PLSS) in preparation for one of the astronauts to leave the LM. All equipment not essential for lunar stay is turned off. The astronauts don their PLSS and depressurize the cabin, open the forward hatch, and exit the vehicle to perform the first of four proposed extravehicular activities (EVA's).

1-4. LUNAR STAY.

During the first EVA, the astronauts activate the modularized equipment stowage assembly (MESA); unstow and deploy the S-band erectable antenna, if required; remove and use the TV, still, and stereo cameras; set the gnomon on the lunar surface; and collect and stow lunar samples. After approximately 4 hours, the first EVA is terminated.

During the second and subsequent EVA's, many of the original activities are repeated and new ones, such as deploying the advanced lunar experiments package (ALSEP), unstowing and deploying the mobility aid, and performing lunar excursions, are initiated.

Upon termination of the final EVA, the astronauts remove their PLSS and jettison all unnecessary equipment to the lunar surface. The LM is then prepared for launch; subsystems are activated and checked and an IMU alignment is performed. At a predetermined launch time, while tracking the CSM with the rendezvous radar, the ascent engine is ignited. The ascent stage of the LM separates from the descent stage and lifts off the lunar surface.

1-5. LIFT-OFF AND TRANSEARTH FLIGHT. (See figure 1-1, sheet 3.)

During the ascent from the lunar surface to the orbital rendezvous with the CSM, the astronauts perform several maneuvers: concentric sequence initiation (CSI), constant delta height maneuver (CDH), terminal phase initiation (TPI), and terminal phase finalization (TPF). At approximately 100 feet from the CSM, all Reaction Control Subsystem (RCS) thrusting is terminated and a CSM-active docking maneuver is performed.

The crew transfers equipment from the LM to the CSM and, after the Commander and the LM Pilot transfer to the CSM, the vehicles are separated and the LM is jettisoned. A brief checkout of the CSM, and determination of transearth thrusting parameters, is followed by the transearth injection maneuver. During the transearth flight, status checks, alignments, and midcourse corrections are performed as required. Approximately 15 minutes before entry into the earth's atmosphere, the SM is jettisoned and the CM is oriented for entry and landing.

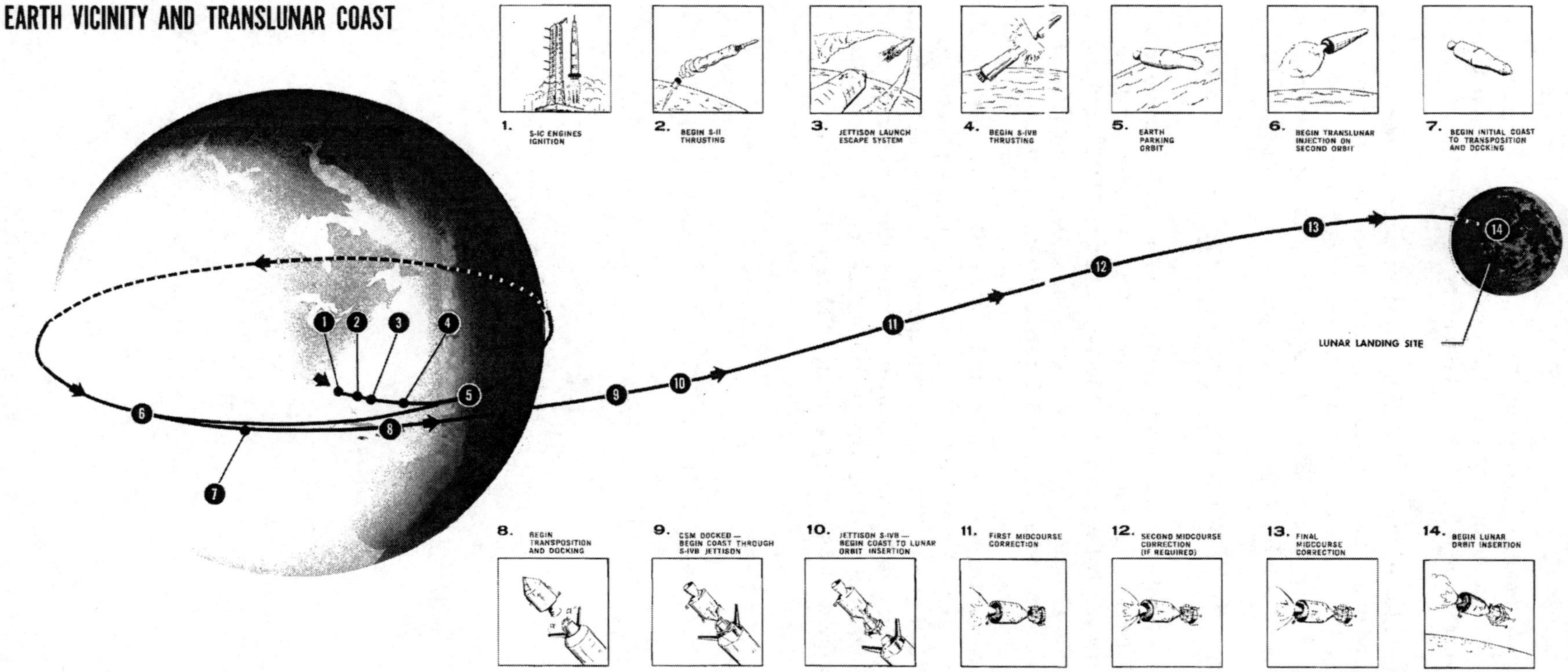

EARTH VICINITY AND TRANSLUNAR COAST

LMA790-2

1. S-IC ENGINES IGNITION
2. BEGIN S-II THRUSTING
3. JETTISON LAUNCH ESCAPE SYSTEM
4. BEGIN S-IVB THRUSTING
5. EARTH PARKING ORBIT
6. BEGIN TRANSLUNAR INJECTION ON SECOND ORBIT
7. BEGIN INITIAL COAST TO TRANSPOSITION AND DOCKING

8. BEGIN TRANSPOSITION AND DOCKING
9. CSM DOCKED — BEGIN COAST THROUGH S-IVB JETTISON
10. JETTISON S-IVB — BEGIN COAST TO LUNAR ORBIT INSERTION
11. FIRST MIDCOURSE CORRECTION
12. SECOND MIDCOURSE CORRECTION (IF REQUIRED)
13. FINAL MIDCOURSE CORRECTION
14. BEGIN LUNAR ORBIT INSERTION

LUNAR LANDING SITE

Figure 1-1. Mission Profile (Sheet 1 of 3)

EARTH VICINITY AND TRANSLUNAR COAST

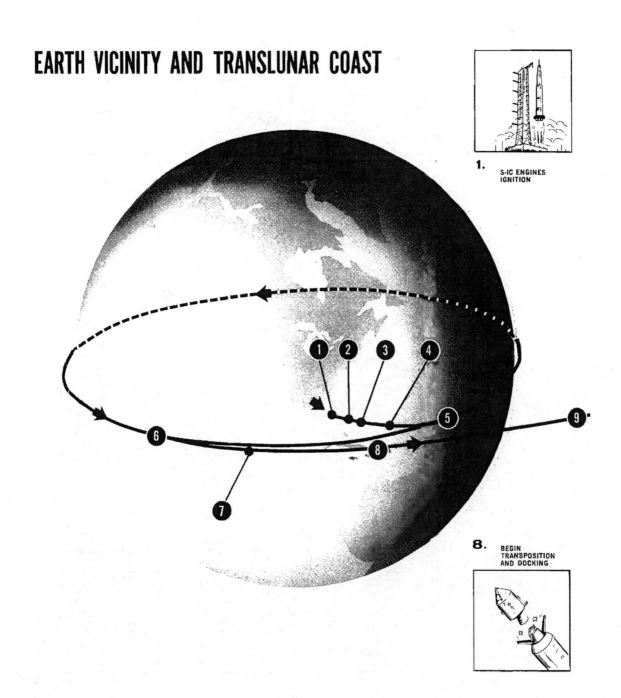

1. S-IC ENGINES IGNITION

8. BEGIN TRANSPOSITION AND DOCKING

2. BEGIN S-II THRUSTING

3. JETTISON LAUNCH ESCAPE SYSTEM

4. BEGIN S-IVB THRUSTING

5. EARTH PARKING ORBIT

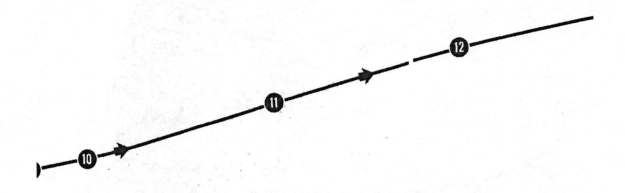

9. CSM DOCKED — BEGIN COAST THROUGH S-IVB JETTISON

10. JETTISON S-IVB — BEGIN COAST TO LUNAR ORBIT INSERTION

11. FIRST MIDCOURSE CORRECTION

12. SECOND MIDCOURSE CORRECTION (IF REQUIRED)

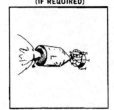

Figure 1

1 November 1969

LMA790-2

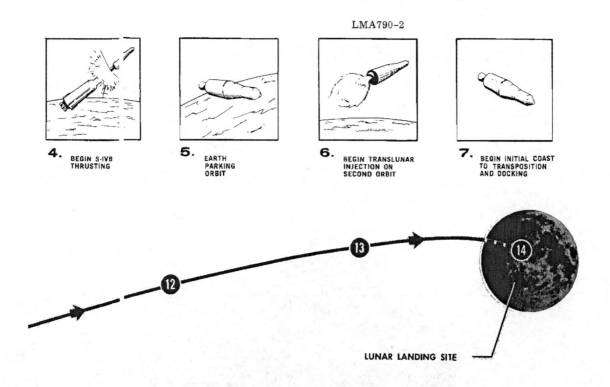

LUNAR LANDING SITE

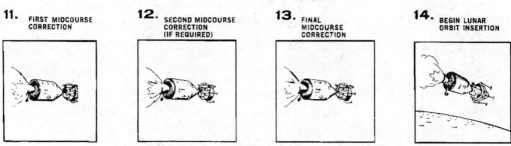

Figure 1-1. Mission Profile (Sheet 1 of 3)

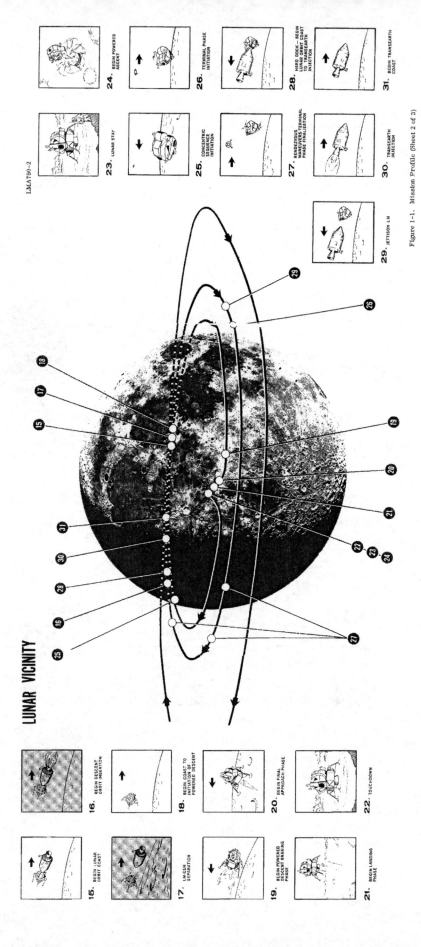

LUNAR VICINITY

LMA790-2

24. BEGIN POWERED ASCENT
26. TERMINAL PHASE INITIATION
28. HARD DOCK -- BEGIN LUNAR ORBIT COAST TO TRANSEARTH INJECTION
31. BEGIN TRANSEARTH COAST

23. LUNAR STAY
25. CONCENTRIC SEQUENCE INITIATION
27. RENDEZVOUS MANEUVERS -- TERMINAL PHASE FINALIZATION
30. TRANSEARTH INJECTION

29. JETTISON LM

Figure 1-1. Mission Profile (Sheet 2 of 3)

16. BEGIN DESCENT ORBIT INSERTION
18. BEGIN COAST TO INITIATION OF POWERED DESCENT
20. BEGIN FINAL APPROACH PHASE
22. TOUCHDOWN

15. BEGIN LUNAR ORBIT COAST
17. LM-CSM SEPARATION
19. BEGIN POWERED DESCENT BRAKING PHASE
21. BEGIN LANDING PHASE

1-7/1-8

1 November 1969

LUNAR VICINITY

15. BEGIN LUNAR ORBIT COAST

16. BEGIN DESCENT ORBIT INSERTION

17. LM-CSM SEPARATION

18. BEGIN COAST TO INITIATION OF POWERED DESCENT

19. BEGIN POWERED DESCENT BRAKING PHASE

20. BEGIN FINAL APPROACH PHASE

21. BEGIN LANDING PHASE

22. TOUCHDOWN

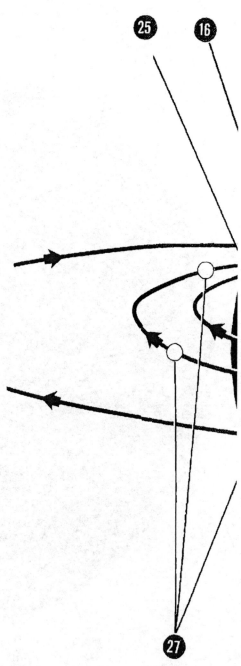

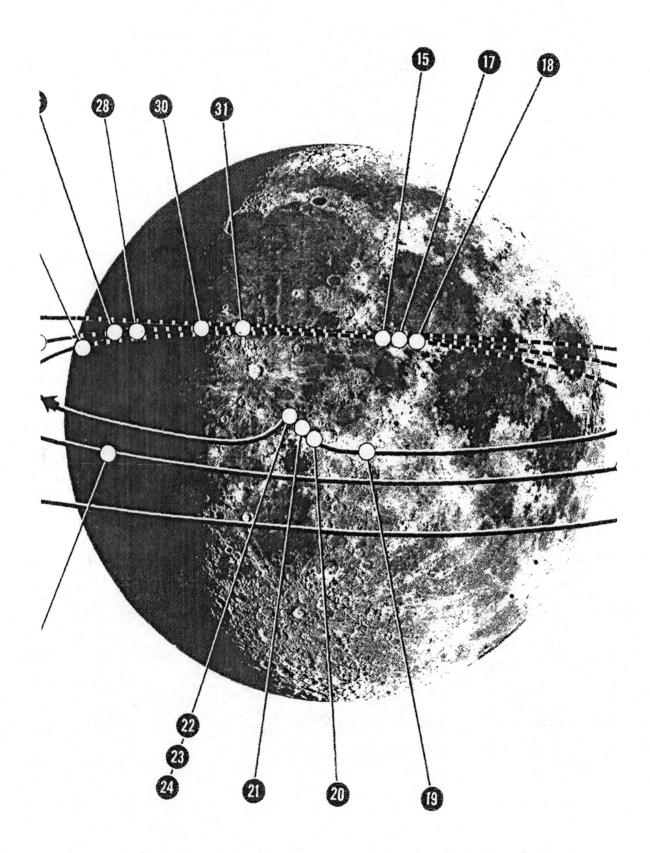

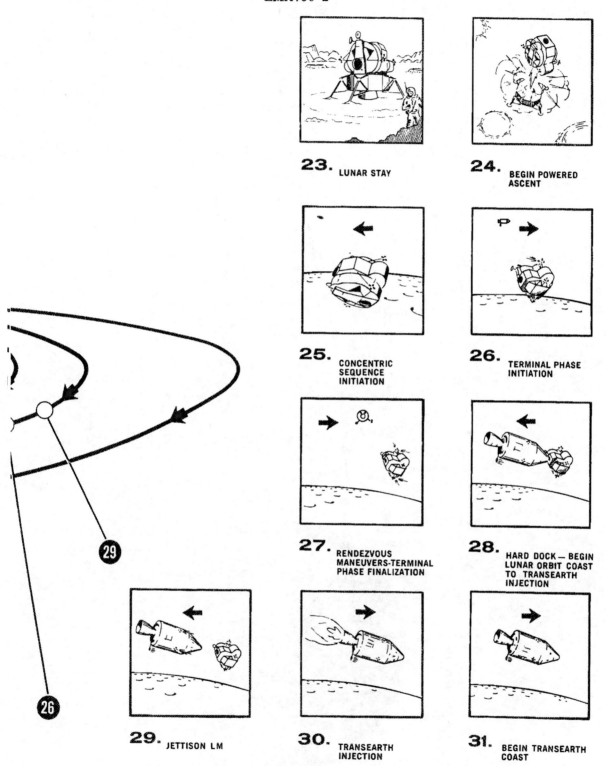

23. LUNAR STAY

24. BEGIN POWERED ASCENT

25. CONCENTRIC SEQUENCE INITIATION

26. TERMINAL PHASE INITIATION

27. RENDEZVOUS MANEUVERS-TERMINAL PHASE FINALIZATION

28. HARD DOCK — BEGIN LUNAR ORBIT COAST TO TRANSEARTH INJECTION

29. JETTISON LM

30. TRANSEARTH INJECTION

31. BEGIN TRANSEARTH COAST

Figure 1-1. Mission Profile (Sheet 2 of 3)

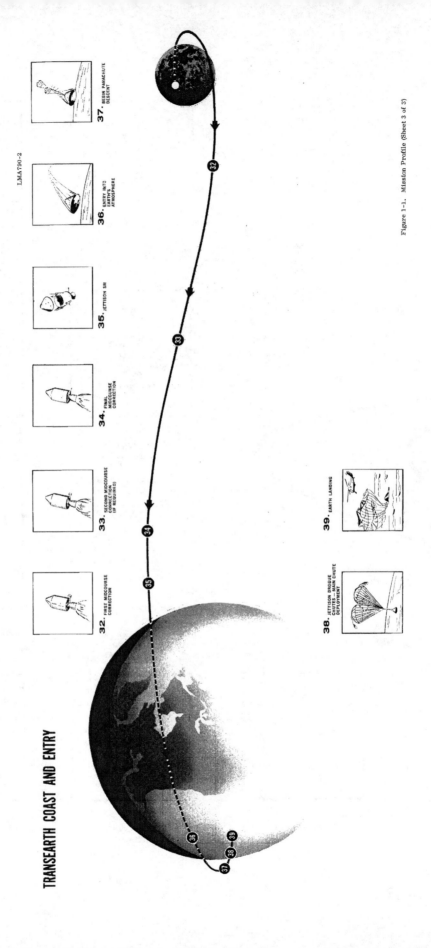

TRANSEARTH COAST AND ENTRY

LMA790-2

32. FIRST MIDCOURSE CORRECTION

33. SECOND MIDCOURSE CORRECTION (IF REQUIRED)

34. FINAL MIDCOURSE CORRECTION

35. JETTISON SM

36. ENTRY INTO EARTH'S ATMOSPHERE

37. BEGIN PARACHUTE DESCENT

38. JETTISON DROGUE CHUTES - MAIN CHUTE DEPLOYMENT

39. EARTH LANDING

Figure 1-1. Mission Profile (Sheet 3 of 3)

TRANSEARTH COAST AND ENTRY

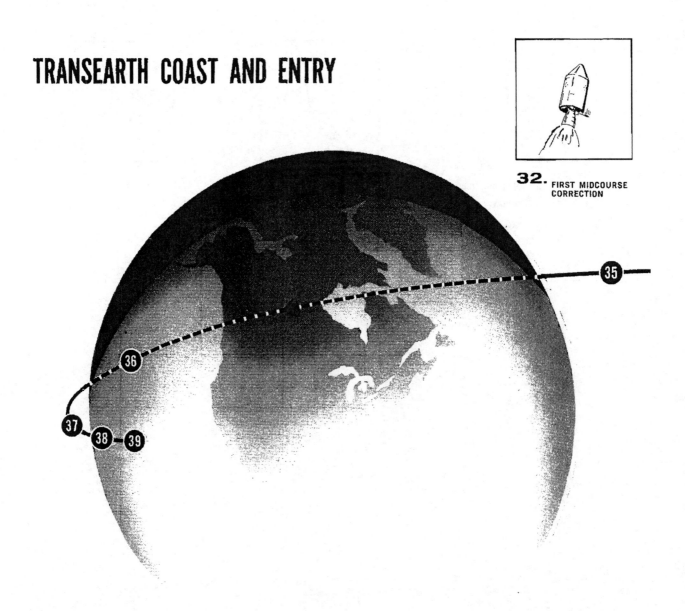

32. FIRST MIDCOURSE CORRECTION

38. JETTISON DROGUE CHUTES — MAIN CHUTE DEPLOYMENT

33. SECOND MIDCOURSE
CORRECTION
(IF REQUIRED)

34. FINAL
MIDCOURSE
CORRECTION

35. JETTISON SM

39. EARTH LANDING

ΓE

Fig

36. ENTRY INTO
EARTH'S
ATMOSPHERE

37. BEGIN PARACHUTE
DESCENT

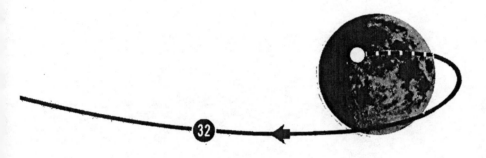

Figure 1-1. Mission Profile (Sheet 3 of 3)

SECTION II

VEHICLE STRUCTURE

2-1. GENERAL. (See figure 2-1.)

The LM consists of an ascent stage and a descent stage. The stages are joined at four interstage fittings by explosive nuts and bolts. Separable interstage umbilicals and hardline connections provide subsystem continuity between the stages. The earth launch weight of the modified vehicle is approximately 36,000 pounds. The overall dimensions of the vehicle are given in figure 2-2.

2-2. ASCENT STAGE. (See figure 2-3.)

The ascent stage is the control center and manned portion of the LM. It comprises three main sections: crew compartment, midsection, and aft equipment bay. The basic structure is primarily aluminum alloy; titanium is used for fittings and fasteners. Aircraft-type construction methods are used. Skin and web panels are chemically milled to reduce weight. Mechanical fasteners join the major structural assemblies, with epoxy as a sealant. Structural members are fusion welded wherever possible, to minimize cabin pressurization leaks. The basic structure includes supports for thrust control engine clusters and antennas. The entire basic structure is enveloped by a thermal and micrometeoroid shield.

2-2.1. CREW COMPARTMENT. (See figure 2-4.)

The crew compartment is the frontal area of the ascent stage; it is cylindrical (92 inches in diameter and 42 inches deep). In this compartment, the crew controls the flight, lunar landing, lunar launch, and rendezvous and docking with the Command/Service Modules. The crew compartment is the operations center during lunar stay.

2-2.1.1. Forward Hatch. The forward hatch, in the front face assembly, is used for transfer of astronauts and equipment between the LM and the lunar surface. A cam latch assembly holds the hatch in the closed position; the assembly forces a lip, around the outer circumference of the hatch, into a elastomeric silicone compound seal that is secured to the vehicle structure. Cabin pressurization forces the hatch lip further into the seal, ensuring a pressure-tight contact. A cabin relief and dump valve is within the hatch structure. A handle is provided on both sides of the hatch, for latch operation. To open the hatch, the cabin must be depressurized.

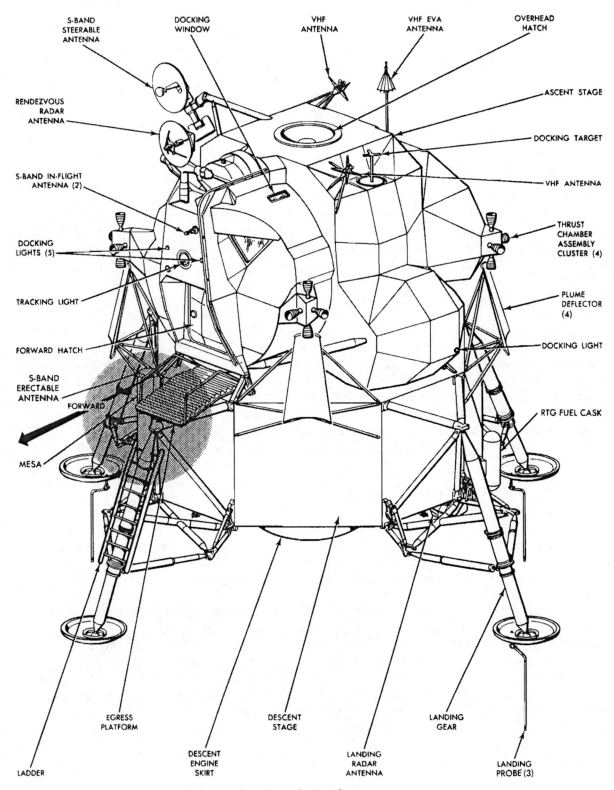

S-BAND
STEERABLE
ANTENNA

DOCKING
WINDOW

VHF
ANTENNA

VHF EVA
ANTENNA

OVERHEAD
HATCH

RENDEZVOUS
RADAR
ANTENNA

ASCENT STAGE

DOCKING TARGET

S-BAND IN-FLIGHT
ANTENNA (2)

VHF ANTENNA

DOCKING
LIGHTS (5)

THRUST
CHAMBER
ASSEMBLY
CLUSTER (4)

TRACKING LIGHT

PLUME
DEFLECTOR
(4)

FORWARD HATCH

DOCKING LIGHT

S-BAND
ERECTABLE
ANTENNA

FORWARD

RTG FUEL CASK

MESA

EGRESS
PLATFORM

DESCENT
STAGE

LANDING
GEAR

LADDER

DESCENT
ENGINE
SKIRT

LANDING
RADAR
ANTENNA

LANDING
PROBE (3)

Figure 2-1. Vehicle Configuration

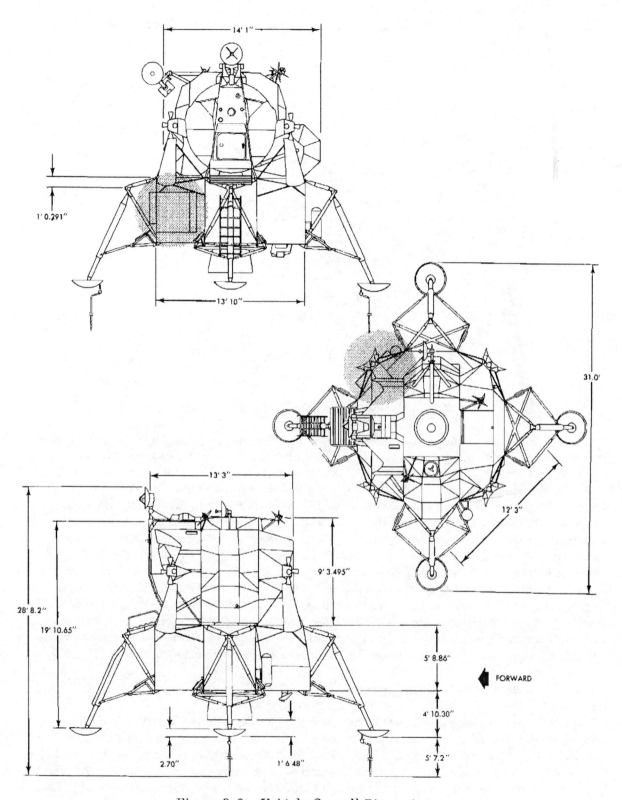

Figure 2-2. Vehicle Overall Dimensions

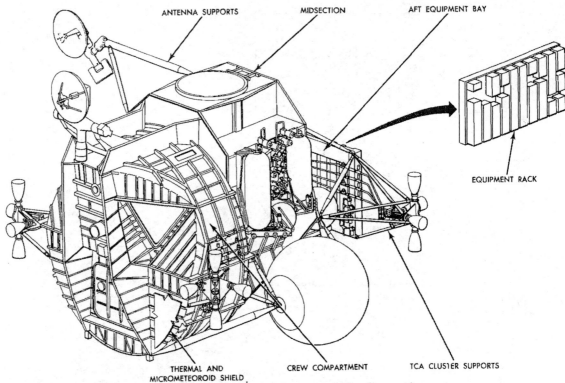

Figure 2-3. Ascent Stage Structure Configuration

2-2.1.2. Windows. Two triangular windows, in the front face assembly, provide visibility during the descent, ascent, and rendezvous and docking phases of the mission. Each window has approximately 2 square feet of viewing area. Both windows are canted down to the side to permit adequate peripheral and downward visibility. A third (docking) window is in the curved overhead portion of the crew compartment, directly above the Commander's flight station. All three windows consist of two separate panes, vented to space environment. The outer pane is of Vycor glass with a thermal (multilayer blue-red) coating on the outboard surface and an antireflective coating on the inboard surface. The antireflective coating is metallic oxide, which reduces the mirror effects of the windows and increases their normal light-transmission efficiency. The inner pane of each window is of chemically tempered, high-strength structural glass. It is sealed with a Raco seal (the docking window inner pane has a dual seal) and has a defog coating on the outboard surface and an antireflective coating on the inboard surface. Both panes are bolted to the window frame through retainers. All three windows are electrically heated to prevent fogging.

2-2.2. MIDSECTION. (See figure 2-5.)

The midsection is immediately aft of the crew compartment. The interior is elliptical, with a minor axis of approximately 56 inches. It is approximately 5 feet high and 54 inches deep. There is a bulkhead at each end. The aft bulkhead supports the aft equipment bay structure. In addition to a lower deck to which the ascent engine is mounted, there are two others. One of these supports the overhead hatch and the lower end of the docking tunnel; the other, supports the upper end of the docking tunnel and absorbs some of the stresses

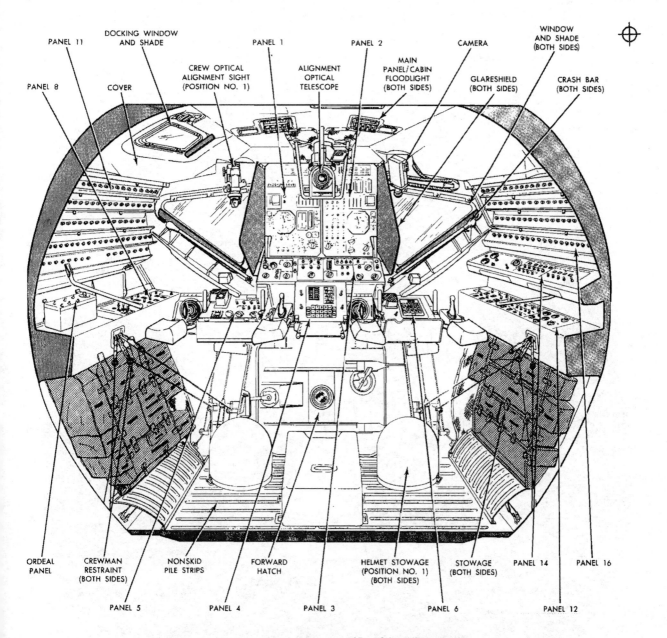

Figure 2-4. Cabin Interior (Looking Forward)

imposed during docking. The right side of the midsection contains most of the Environmental Control Subsystem (ECS) controls and most of the heat transport section water-glycol plumbing. Valves for operation of the ECS equipment are readily accessible from the crew compartment. The left side of the midsection contains the waste management section, a portable life support system (PLSS), and other crew provision stowage. Guidance, Navigation, and Control Subsystem (GN&CS) electronic units that do not require access by the astronauts are on the midsection aft bulkhead. Reaction Control Subsystem (RCS) propellant tanks are installed between the midsection bulkheads, on each side, external to the basic structure of the midsection. The ascent engine propellant tanks are mounted in the midsection, beneath the RCS tanks.

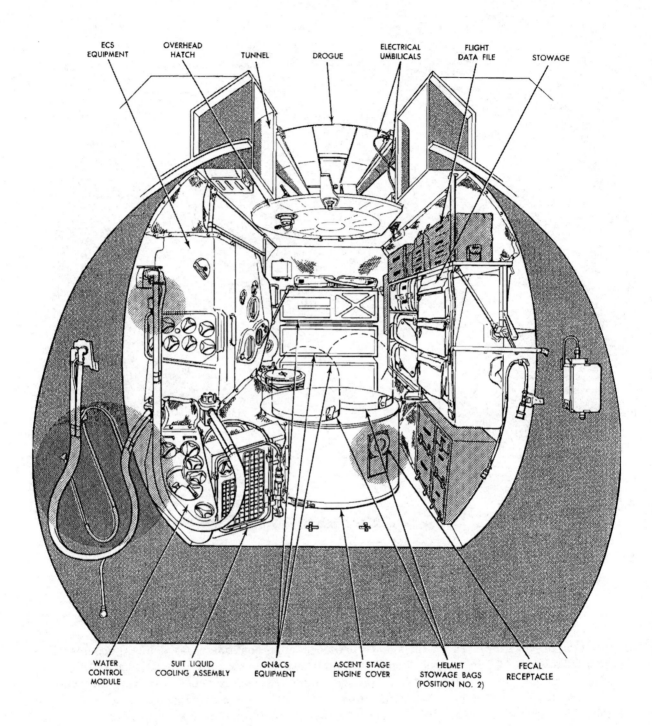

Figure 2-5. Cabin Interior (Looking Aft)

2-2.2.1. <u>Overhead Hatch.</u> The overhead hatch, approximately 33 inches in diameter, is directly above the ascent engine cover, at the top centerline of the midsection. The hatch opens inward and permits transfer of astronauts and equipment when the LM and the CM are docked. An off-center latch adjacent to the forward edge of the hatch, can be operated from either side of the hatch. A maximum torque of 35 inch-pounds is required to disengage the latching mechanism to open the hatch. A elastomeric silicone compound seal is mounted in the hatch frame structure. When the latch is closed, a lip near the outer circumference of the hatch enters the seal, ensuring a pressure-tight contact. A cabin relief and dump valve is within the hatch structure. Normal cabin pressurization forces the hatch into its seal. To open the hatch, the cabin must be depressurized.

2-2.2.2. <u>Docking Tunnel.</u> The docking tunnel, immediately above the overhead hatch, provides a structural interface between the LM and the CM to permit transfer of equipment and astronauts without exposure to space environment. The tunnel is 33 inches in diameter and 18 inches long. The lower end of the tunnel is welded to the upper deck structure; the upper end is secured to the main beams and the outer deck.

2-2.3. AFT EQUIPMENT BAY.

The aft equipment bay is separated from the midsection by a pressure-tight bulkhead; the bay is unpressurized. The main supporting structure of the bay consists of tubular truss members bolted to the aft side of the bulkhead. The truss members, used in a cantilever type of construction, extend aft to the equipment rack. The equipment rack is constructed of a series of vertical box beams, supported by an upper and lower Z-frame. The beams have integral cold rails that transfer heat from the electronic equipment mounted on the equipment rack. The cold rails are mounted vertically in the structural frame, which is supported at its upper and lower edges by the truss members. A water-gylcol solution (coolant) flows through the cold rails.

2-2.4. THERMAL AND MICROMETEOROID SHIELD.

The ascent stage thermal and micrometeoroid shield combines either a blanket of multiple layers of aluminized polyimide sheet (Kapton H-film) and aluminized polyester sheet (mylar) with a sandwich of inconel mesh and nickel foil or a polyimide blanket with a single sheet of aluminum skin. The combined thermal and micrometeoroid shield is mounted on supports (standoffs), which keep it at least 2 inches from the main structure. The standoffs have low thermal conductivity. Where subsystem components are mounted external to the ascent stage basic structure, the standoffs are mounted to aluminum frames that surround the components. The aluminum or inconel (the outermost material) serves as a micrometeoroid bumper; the sandwich and blanket material serve as thermal shielding. Where the blankets meet, the mating edges are sealed with mylar tape. The blankets have vent holes. During earth prelaunch activities, various components and areas of the ascent stage must be readily accessible. Access panels in the outer skin and insulation provide this accessibility.

2-3.   DESCENT STAGE.   (See figure 2-6.)

The descent stage is the unmanned portion of the LM. The descent stage structure of
aluminum-alloy, chemically milled webs provides attachment and support points for securing the LM within the spacecraft - Lunar Module adapter (SLA).   The structure consists
of two pair of parallel beams arranged in a cruciform, with a deck on the upper and lower
surfaces approximately 65 inches apart. The ends of the beams are closed off by end closure
bulkheads to provide five equally sized compartments: a center compartment, one forward
and one aft of the center compartment, and one right and one left of the center compartment.
The center compartment houses the descent engine. Descent engine oxidizer tanks are
housed in the forward and aft compartments; descent engine fuel tanks, in the side compartments. The entire basic structure is enveloped by a thermal and micrometeoroid shield.

The areas, between the compartments, that give the descent stage its octagonal shape are
referred to as quadrants. The quadrants are designated 1 through 4, beginning at the left
of the forward compartment and continuing counterclockwise (as viewed from the top)
around the center. Quadrant 1 houses the payload of the vehicle. Quadrant 2 houses an
ECS water tank, and a modified Apollo lunar scientific equipment package (ALSEP). A fuel
cask for use with the radioisotope thermoelectric generator (RTG) is mounted adjacent to the
ALSEP, but outside the quadrant thermal blanket and micrometeoroid shield. A landing
radar antenna is supported externally on additional structure below the lower deck. Quadrant
3 houses supercritical helium and ambient helium tanks, the descent engine control assembly
of the GN&CS, and a gaseous oxygen tank. Quadrant 4 houses Explosive Devices Subsystem
(EDS) components, a waste management container, and a modified modularized equipment
stowage assembly (MESA). A descent water tank and gaseous exygen tank, and their supporting structure, have been added to quadrant 4. The S-band antenna housed in quadrant 1
of the LM is housed (attached to the MESA) in quadrant 4 of the modified vehicle.

Five EPS batteries and two electrical control assemblies (ECA's) are mounted on the rear
bulkhead (-Z) of the aft compartment. Four plume deflectors truss mounted to the descent
stage divert the plume of the downward-firing RCS thrusters away from the descent stage.
A landing gear attenuates landing impact and supports the vehicle. The deflectors in Quadrants I and IV have been shortened to provide the necessary clearance for the payload and
MESA, respectively. The supporting trusses have also been modified.

2-3.1. THERMAL AND MICROMETEOROID SHIELD.

The descent stage thermal shield combines multiple layers of aluminized mylar and H-film
with an outer skin of H-film. In areas where micrometeorite protection is required, one layer
of black-painted inconel is used as skin. The shield is mounted on supports, which keep it
at least 1/2 inch away from the main structure. The supports have low thermal conductivity.
A base heat shield, composed of titanium with a blanket of alternate layers of nickel foil and
fiberfax outside, protects the bottom of the descent stage from engine heat. In addition, the
engine compartment is protected by a titanium shield with a thermal blanket of multiple
layers of nickel foil and fiberfax under an outer blanket of H-film.

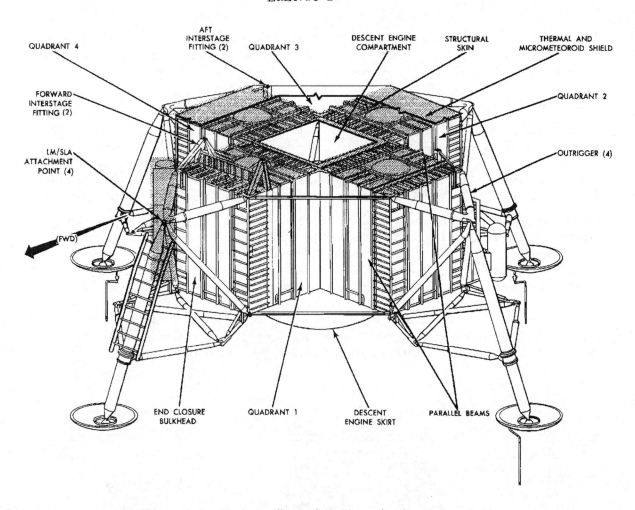

Figure 2-6. Descent Stage Structure Configuration

2-3.2. LANDING GEAR.

The landing gear is of the cantilever type; it consists of four assemblies, each connected to an outrigger that extends from the ends of the structural parallel beams. The landing gear assemblies extend from the front, rear, and both sides of the descent stage. Each assembly consists of struts, trusses, a footpad, and lock and deployment mechanisms. The left, right, and aft footpad has a lunar surface sensing probe. A ladder is affixed to the forward gear assembly.

The landing gear attenuates the impact of a lunar landing, prevents tipover, and supports the vehicle during lunar stay and lunar launch. Compression loads are attenuated by a crushable aluminum-honeycomb cartridge in each primary strut. Landing impact is attenuated to load levels that preserves the vehicle structural integrity. At earth launch, the landing gear is in the retracted condition. When the Commander, in the vehicle, operates the landing gear deployment switch, the landing gear uplocks are explosively released and springs in the deployment mechanism extend the landing gear. Once extended, each gear assembly is locked in place by two downlock mechanisms. The lunar surface sensing probe is an electromechanical device. The probes are retained in the stowed position against the primary strut until landing gear deployment. During deployment, mechanical interlocks are released, permitting spring energy to extend the probes below the footpad. At lunar contact, two mechanically actuated switches in each probe energize lights to advise the crew to shut off the descent engine.

2-4.   INTERSTAGE ATTACHMENTS AND SEPARATIONS.   (See figure 2-7.)

At earth launch, the LM is housed within the SLA, which has an upper and lower section. The outriggers, to which the landing gear is attached, provide attachment points for securing the vehicle to the SLA lower section. The SLA upper panels are deployed and jettisoned when the CSM is separated from the SLA. These panels, which are hinged to the lower section, fold back and are then forced away from the SLA by spring thrusters.

After transposition, the CSM docks with the LM. A ring at the top of the ascent stage provides a structural interface for joining the LM to the CM. The ring is compatible with the clamping mechanism in the CM and provides structural continuity. The drogue portion of the docking mechanism is secured below this ring. The drogue is required during docking operations to mate with the CM-mounted probe.

When docking has been completed, the astronauts connect electrical umbilicals in the CM and the LM to provide electrical power to the LM for separation from the SLA. The vehicle is then explosively separated from the SLA lower section

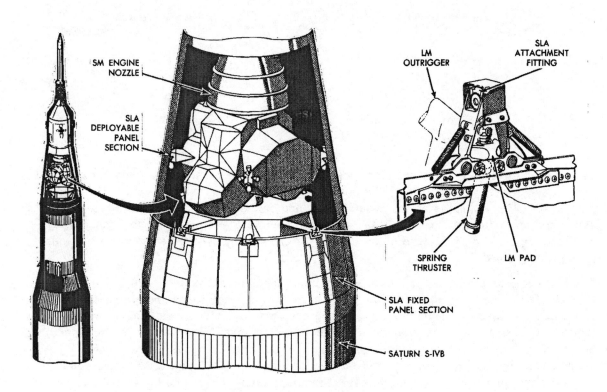

Figure 2-7.   LM Interface With SLA

SECTION III

OPERATIONAL SUBSYSTEMS

3-1. GENERAL.

This section describes the LM operational subsystems in sufficient detail to convey an understanding of the LM as an integrated system. The integrated LM system comprises the following subsystems:

- Guidance, Navigation, and Control
- Reaction Control
- Propulsion
- Instrumentation
- Controls and Displays
- Radar Subsystem

- Communications
- Electrical Power
- Environmental Control
- Crew Personal Equipment
- Explosive Devices Subsystem
- Lighting

3-2. GUIDANCE, NAVIGATION, AND CONTROL SUBSYSTEM.

The primary function of the Guidance, Navigation, and Control Subsystem (GN&CS) is accumulation, analysis, and processing of data to ensure that the vehicle follows a predetermined flight plan. The GN&CS provides navigation, guidance, and flight control to accomplish the specific guidance goal. To accomplish guidance, navigation, and control, the astronauts use controls and indicators that interface with the various GN&CS equipment. Functionally, this equipment is contained in a primary guidance and navigation section (PGNS), an abort guidance section (AGS), and a control electronics section (CES). (See figure 3-2.1.)

The PGNS provides the primary means for implementing inertial guidance and optical navigation for the vehicle. When aided by either the rendezvous radar (RR) or the landing radar (LR), the PGNS provides for radar navigation. The section, when used in conjunction with the CES, provides automatic flight control. The astronauts can supplement or override automatic control, with manual inputs.

The PGNS acts as a digital autopilot in controlling the vehicle throughout the mission. Normal guidance requirements include transferring the vehicle from a lunar orbit to its descent profile, achieving a successful landing at a preselected or crew-selected site, and performing a powered ascent maneuver that results in terminal rendezvous with the CSM.

The PGNS provides the navigational data required for vehicle guidance. These data include line-of-sight (LOS) data from an alignment optical telescope (AOT) for inertial reference alignment, signals for initializing and aligning the AGS, and data to the astronauts for determining the location of the computed landing site.

The AGS is primarily used only if the PGNS malfunctions. If the PGNS is functioning properly when a mission is aborted, it is used to control the vehicle. Should the PGNS fail, the lunar mission would have to be aborted; thus, the term "abort guidance section." Abort guidance provides only guidance to place the vehicle in a rendezvous trajectory with the CSM or in a parking orbit for CSM-active rendezvous. The navigation function is performed by the PGNS, but the navigation information also is supplied to the AGS. In case of a PGNS malfunction, the AGS uses the last navigation data provided to it. The astronaut can update the navigation data by manually inserting RR data into the AGS.

The AGS is used as backup for the PGNS during a mission abort. It determines the vehicle trajectory or trajectories required for rendezvous with the CSM and can guide the vehicle from any point in the mission, from separation to rendezvous and docking, including ascent from the lunar surface. It can provide data for attitude displays, make explicit guidance computations, and issue commands for firing and shutting down the engines. Guidance can be accomplished automatically, or manually by the astronauts, based on data from the AGS. When the AGS is used in conjunction with the CES, it functions as an analog autopilot.

The AGS is an inertial system that is rigidly strapped to the vehicle rather than mounted on a stabilized platform. Use of the strapped-down inertial system, rather than a gimbaled system, offers sufficient accuracy for lunar missions, with savings in size and weight. Another feature is that it can be updated manually with radar and optical aids.

The CES processes Reaction Control Subsystem (RCS) and Main Propulsion Subsystem (MPS) control signals for vehicle stabilization and control. To stabilize the vehicle during all phases of the mission, the CES provides signals that fire any combination of the 16 RCS thrusters. These signals control attitude and translation about or along all axes. The attitude and translation control data inputs originate from the PGNS during normal automatic operation, from two hand controllers during manual operations, or from the AGS during certain abort situations.

The CES also processes on and off commands for the ascent and descent engines and routes automatic and manual throttle commands to the descent engine. Trim control of the gimbaled descent engine is also provided to assure that the thrust vector operates through the vehicle center of gravity.

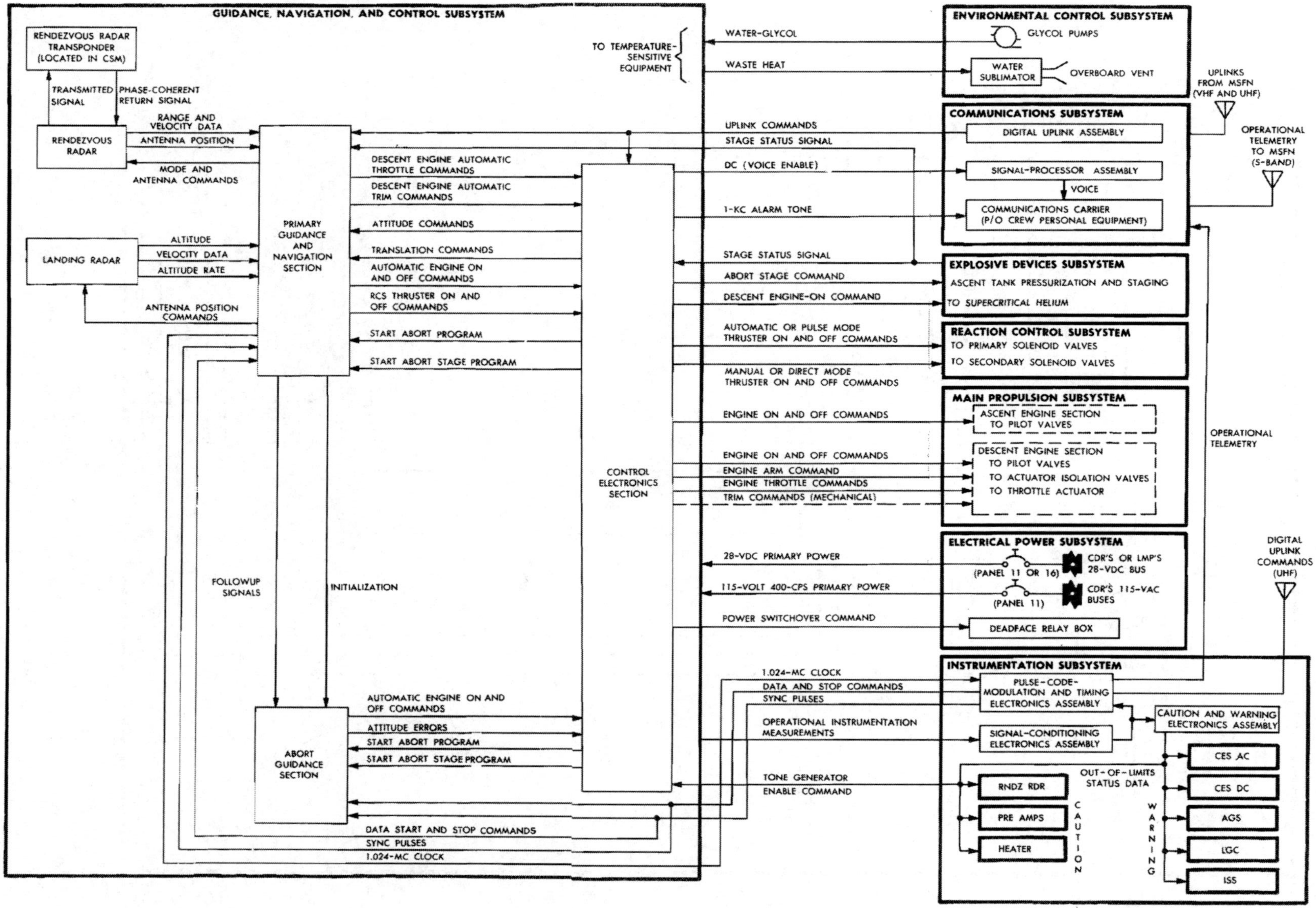

Figure 3-2.1. Guidance, Navigation, and Control Subsystem – Simplified Block Diagram and Subsystem Interfaces

1 November 1969

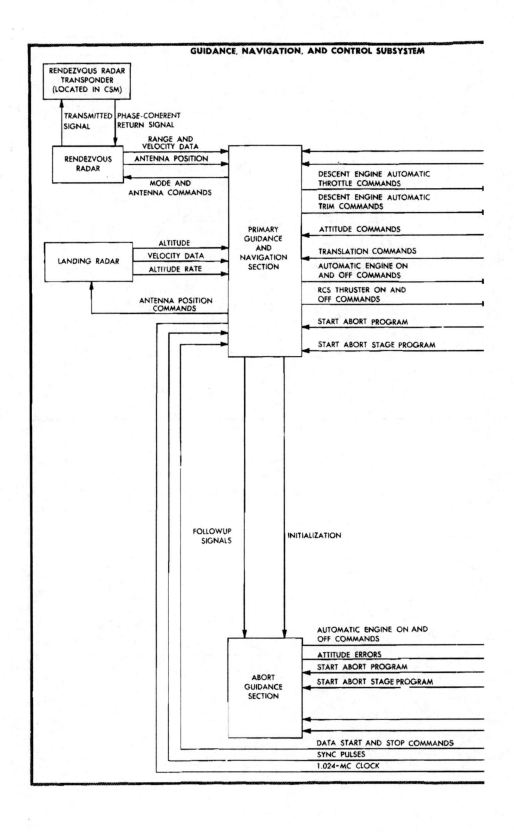

GUIDANCE, NAVIGATION, AND CONTROL SUBSYSTEM

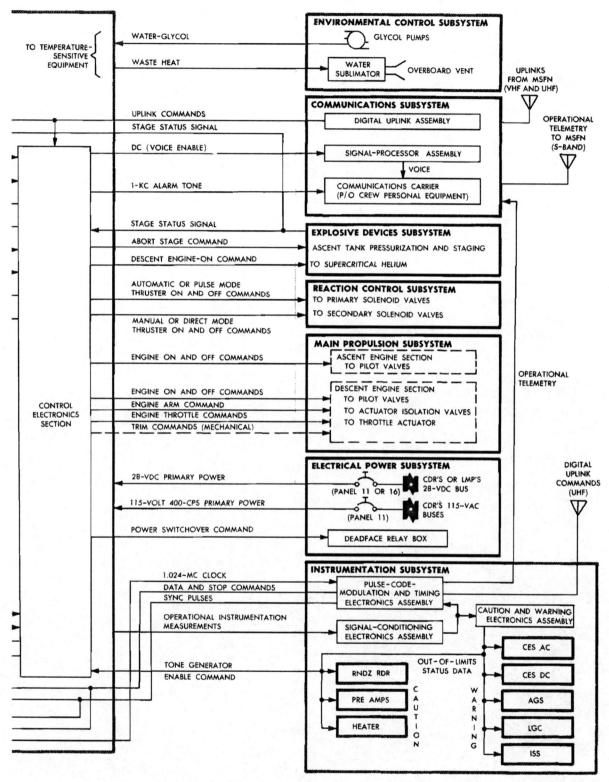

Figure 3-2.1. Guidance, Navigation, and Control Subsystem -
Simplified Block Diagram and Subsystem Interfaces

These integrated sections (PGNS, AGS, and CES) allow the astronauts to operate the vehicle in fully automatic, several semiautomatic, and manual control modes.

3-2.1 PRIMARY GUIDANCE AND NAVIGATION SECTION.

The PGNS includes three major subsections: inertial, optical, and computer. (See figure 3-2.2.) Individually or in combination they perform all the functions mentioned previously.

The inertial subsection (ISS) establishes the inertial reference frame that is used as the central coordinate system from which all measurements and computations are made. The ISS measures attitude and incremental velocity changed, and assists in converting data for computer use, onboard display, or telemetry. Operation is started automatically by a guidance computer or by an astronaut using the computer keyboard. Once the ISS is energized and aligned to the inertial reference, any vehicle rotation (attitude change) is sensed by a stable platform. All inertial measurements (velocity and attitude) are with respect to the stable platform. These data are used by the computer in determining solutions to the guidance problems. The ISS consists of a navigation base, an inertial measurement unit (IMU), a coupling data unit (CDU), pulse torque assembly (PTA), power and servo assembly (PSA), and signal conditioner assembly (SCA). (See figure 3-2.3.)

The optical subsection (OSS) is used to determine the position of the vehicle using a catalog of stars stored in the computer and celestial measurements made by an astronaut. The identity of celestial objects is determined before earth launch. The AOT is used by the astronaut to take direct visual sightings and precise angular measurements of a pair of celestial objects. The computer subsection (CSS) uses this data, along with prestored data, to compute position and velocity and to align the inertial components. The OSS consists of the AOT and a computer control and reticle dimmer (CCRD) assembly. (See figure 3-2.2.)

The CSS, as the control and data-processing center of the vehicle, performs all the guidance and navigation functions necessary for automatic control of the flight path and attitude of the vehicle. For these functions, the GN&CS uses a digital computer. The computer is a control computer with many of the features of a general-purpose computer. As a control computer, it aligns the stable platform, and positions both radar antennas. It also provides control commands to both radars, the ascent engine, the descent engine, the RCS thrusters, and the vehicle cabin displays. As a general-purpose computer, it solves guidance problems required for the mission. The CSS consists of a LM guidance computer (LGC) and a display and keyboard (DSKY), which is a computer control panel. (See figure 3-2.3.)

3-2.1.1. Navigation Base. The navigation base is a lightweight (approximately 3 pounds) mount that supports, in accurate alignment, the IMU, AOT, and an abort sensor assembly (ASA). Structurally, it consists of a center ring with four legs that extend from either side of the ring. The IMU is mounted to the legs on one end; the AOT and the ASA are mounted on the opposite side.

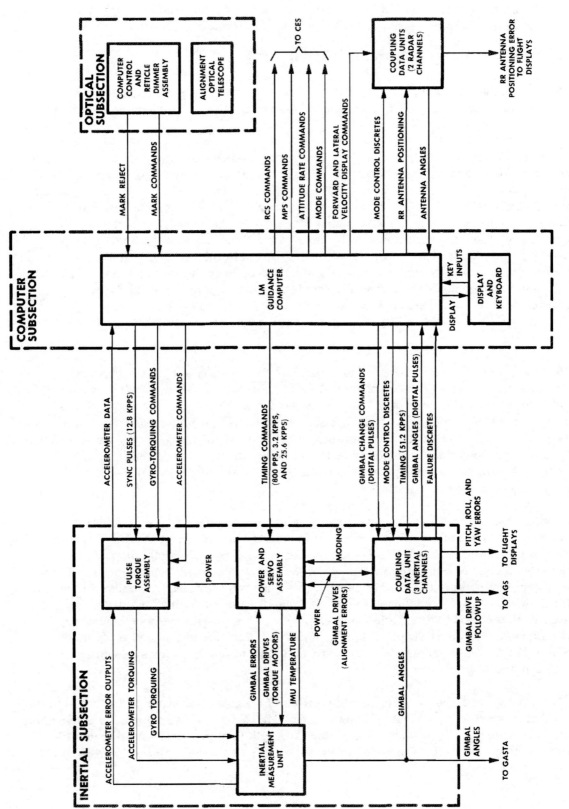

Figure 3-2.2. Primary Guidance and Navigation Section – Block Diagram

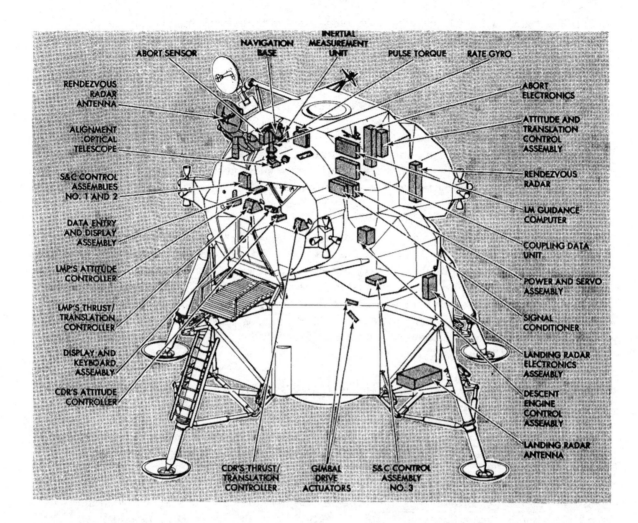

Figure 3-2.3. Guidance, Navigation, and Control Subsystem-Major Equipment Location

3-2.1.2. <u>Inertial Measurement Unit</u>. The IMU is the primary inertial sensing device of the vehicle. It is three-degree-of-freedom, stabilized device that maintains an orthogonal, inertially referenced coordinate system for vehicle attitude control and maintains three accelerometers in the reference coordinate system for accurate measurement of velocity changes. The IMU contains a stable platform, gyroscopes, and accelerometers necessary to establish the inertial reference.

The stable platform serves as the space-fixed reference for the ISS. It is supported by three gimbal rings (outer, middle, and inner) for complete freedom of motion. Three Apollo inertial reference integrating gyroscopes sense attitude changes; they are mounted on the stable platform, mutually perpendicular. The gyros are fluid- and magnetically-suspended, single-degree-of-freedom types. They sense displacement of the stable platform and generate error signals proportional to displacement. Three pulse integrating pendulous accelerometers (fluid- and magnetically-suspended devices) sense velocity changes.

3-2.1.3. <u>Coupling Data Unit.</u> The CDU converts and transfers angular information between the GN&CS hardware. The unit is an electronic device that performs analog-to-digital and digital-to-analog conversion. The CDU processes the three attitude angles associated with the inertial reference and the two angles associated with the RR antenna. It consists of five almost identical channels: one each for the inner, middle, and outer gimbals of the IMU and one each for the RR shaft and trunnion gimbals.

The two channels used with the RR interface between the RR antenna and the LGC. The LGC calculates digital antenna position commands before acquisition of the CSM. These signals, converted to analog form by the CDU, are applied to the antenna drive mechanism to aim the antenna. Analog tracking-angle information, converted to digital form by the unit, is applied to the LGC.

The three channels used with the IMU provide interfaces between the IMU and the LGC and between the LSC and the AGS. Each of the three IMU gimbal angle resolvers provides its channel with analog gimbal-angle signals that represent vehicle attitude. The CDU converts these signals to digital form and applies them to the LGC. The LGC calculates attitude or translation commands and routes them through the CES to the proper thruster. The CDU converts attitude error signals to 800-Hz analog signals and applies them to the FDAI. Coarse- and fine-alignment commands generated by the LGC are coupled to the IMU through the CDU.

3-2.1.4. <u>Pulse Torque Assembly.</u> The PTA supplies inputs to, and processes outputs from, the inertial components in the ISS.

3-2.1.5. <u>Power and Servo Assembly.</u> The PSA contains power supplies for generation of internal power required by the PGNS, and servomechanisms and temperature control circuitry for the IMU.

3-2.1.6. <u>Signal Conditioner Assembly.</u> The SCA provides an interface between the PGNS and the Instrumentation Subsystem (IS). The SCA preconditions PGNS measurements to a 0- to 5-volt d-c format before the signals are routed to the IS.

3-2.1.7. <u>Alignment Optical Telescope.</u> The AOT, an L-shaped periscope approximately 36 inches long, is used by the astronaut to take angular measurements of celestial objects. These angular measurements are required for orienting the platform during certain periods while the vehicle is in flight and during prelaunch preparations while on the lunar surface. Sightings taken with the AOT are transferred to the LGC by the astronaut, using the CCRD assembly. This assembly also controls the brightness of the telescope reticle pattern.

The AOT is a unity-power, periscope-type device with a 60° conical field of view. It has a movable shaft axis (parallel to the LM X-axis) and a LOS approximately 45° from the X-axis in the Y-Z plane. The LOS is fixed in elevation and movable in azimuth to six detent positions at 60° intervals. Detent positions are selected by turning a selector knob on the AOT.

The reticle pattern within the eyepiece optics consists of crosshairs and a pair of Archimedes spirals. The vertical crosshair, an orientation line designated the Y-line, is parallel to the X-axis when the reticle is at the 0° reference position. The horizontal crosshair, an auxiliary line designated the X-line, is perpendicular to the orientation line. The one-turn spirals are superimposed from the center of the field of view to the top of the vertical crosshair. Ten miniature red lamps mounted around the reticle prevent false star indications caused by imperfections in the reticle and illuminate the reticle pattern. Stars will appear white; reticle imperfections, red. Heaters prevent fogging of the mirror due to moisture and low temperatures during the mission.

A rotable eyeguard, fastened to the end of eyepiece, is axially adjustable for head position. The eyeguard is used when the astronaut takes sightings with his faceplate open. It is removed when the astronaut takes sightings with his faceplate closed; a fixed eyeguard, permanently cemented to the AOT is used instead. The fixed eyeguard prevents marring of the faceplate by the eyepiece. A high-density filter lens, supplied as auxiliary equipment, prevents damage to the astronaut's eyes due to accidental direct viewing of the sun or if the astronaut chooses to use the sun as a reference.

3-2.1.8. <u>Computer Control and Reticle Dimmer Assembly.</u> The CCRD assembly is mounted on an AOT guard. The mark X and mark Y pushbuttons are used by the astronauts to send discrete signals to the LGC when star sightings are made. The reject pushbutton is used if an invalid mark has been sent to the LGC. A thumbwheel on the assembly adjusts the brightness of the telescope reticle lamps.

3-2.1.9. <u>LM Guidance Computer.</u> The LGC is the central data-processing device of the GN&CS. The LGC, a control computer with many of the features of a general-purpose computer, processes data and issues discrete control signals for various subsystems. As a control computer, it aligns the IMU stable platform and provides RR antenna drive commands. The LGC also provides control commands to the LR and RR, the ascent and descent engines, the RCS thrusters, and the cabin displays. As a general-purpose computer, it solves guidance problems required for the mission. In addition, the LGC monitors the operation of the PGNS.

The LGC stores data pertinent to the ascent and descent flight profiles that the vehicle must assume to complete its mission. These data (position, velocity, and trajectory information) are used by the LGC to solve flight equations. The results of various equations are used to determine the required magnitude and direction of thrust. The LGC establishes corrections to be made. The vehicle engines are turned on at the correct time, and steering commands are controlled by the LGC to orient the vehicle to a new trajectory, if required. The ISS senses acceleration and supplies velocity changes, to the LGC, for calculating total velocity. Drive signals are supplied from the LGC to the CDU and stabilization gyros in the ISS to align the gimbal angles in the IMU position signals are supplied to the LGC to indicate attitude changes.

The LGC provides antenna-positioning signals to the RR and receives, from the RR channels of the CDU, antenna angle information. The LGC uses this information in the antenna-positioning calculations. During lunar-landing operations, star-sighting information is manually loaded into the LGC, using the DSKY. This information is used to calculate IMU alignment commands. The LGC and its programming help meet the functional requirements of the mission. The functions performed in the various mission phases include automatic and semiautomatic operations that are implemented mostly through the execution of the programs stored in the LGC memory.

The LGC is a parallel fixed-point, one's complement, digital computer with a fixed rope core memory and an erasable ferrite-core memory. It has a limited self-check capability. Inputs to the LGC are received from the LR and RR, from the IMU through the inertial channels of the CDU, and from an astronaut through the DSKY. The LGC memory consists of an erasable and a fixed magnetic core memory with a combined capacity of 38,916 16-bit words. The erasable memory is a coincident-current, ferrite core array with a total capacity of 2,048 words; it is characterized by destructive readout. The fixed memory consists of three magnetic-core rope modules. Each module contains two sections; each section contains 512 magnetic cores. The capacity of each core is 12 words, making a total of 36,864 words in the fixed memory. Readout from the fixed memory is non-destructive.

The LGC performs all necessary arithmetic operations by addition, adding two complete words and preparing for the next operation in approximately 24 microseconds. To subtract, the LGC adds the complement of the subtrahend. Multiplication is performed by successive additions and shifting; division, by successive addition of complements and shifting.

Functionally, the LGC contains a timer, sequence generator, central processor, priority control, an input-output section, and a memory unit. The timer generates all necessary synchronization pulses to ensure a logical data flow with the subsystems. The sequence generator directs the execution of the programs. The central processor performs all arithmetic operations and checks information to and from the LGC. Memory stores the LGC data and instructions. Priority control establishes a processing priority for operations that must be performed by the LGC. The input-output section routes and conditions signals between the LGC and the other subsystems.

The main functions of the LGC are implemented through execution of programs stored in memory. Programs are written in machine language called basic instructions. A basic instruction can be an instruction word or a data word. Instruction words contain a 12-bit address code and a three-bit order code. The LGC operates in an environment in which many parameters and conditions change in a continuous manner. The LGC however, operates in an incremental manner, one item at a time. Therefore, for it to process the parameters, its hardware is time shared. The time sharing is accomplished by assigning priorities to the processing functions. These priorities are used by the LGC so that it processes the highest priority processing function first.

3-2.1.10. <u>Display and Keyboard</u>. Through the DSKY, the astronaut can load information into the LGC, retrieve and display information contained in the LGC, and initiate any program stored in memory. The astronauts can also use the DSKY to control the moding of the ISS. The exchange of data between the astronauts and the LGC is usually initiated by an astronaut; however, it can also be initiated by internal computer programs.

The DSKY is located on panel 4, between the Commander and LM Pilot and above the forward hatch. The upper half is the display portion; the lower half comprises the keyboard. The display portion contains five caution indicators, six status indicators, seven operation display indicators, and three data display indicators. These displays provide visual indications of data being loaded in the LGC, the condition of the LGC, and the program being used. The displays also provide the LGC with a means of displaying or requesting data.

3-2.2. ABORT GUIDANCE SECTION. (See figure 3-2.5.)

The AGS consists of an abort sensor assembly (ASA), abort electronics assembly (AEA), and a data entry and display assembly (DEDA). The ASA performs the same function as the IMU; it establishes an inertial reference frame. The AEA, a high-speed general-purpose digital computer is the central processing and computational device for the AGS. The DEDA is the input-output device for controlling the AEA.

3-2.2.1. Abort Sensor Assembly. The ASA, by means of gyros and accelerometers, provides incremental attitude information around the vehicle X, Y, and Z axes and incremental velocity changes along the vehicle X, Y, and Z axes. Data pulses are routed to the AEA which uses the attitude and velocity data for computation of steering errors.

The ASA consists of three strapped-down pendulous accelerometers, three strapped-down gyros, and associated electronic circuitry. The accelerometers and gyros (one each for each vehicle axis) sense body-axis motion with respect to inertial space. The accelerometers sense acceleration along the vehicle orthogonal axis. The gyros and accelerometers are securely fastened to the vehicle X, Y, and Z axes so that motion along or around one or more axis is sensed by one or more gyros or accelerometers.

The strapped-down inertial guidance system has the advantage of substantial size and weight reduction over the more conventional gimbaled inertial guidance system, but has the disadvantage of error buildup over sustained periods of operation. Calibration uses the PGNS as a reference to determine the drift-compensation parameters for the ASA gyros. Calibration parameters stored in the AEA are used to correct calculations based on the gyro inputs.

3-2.2.2. Data Entry and Display Assembly. The DEDA is used by the astronauts to select the desired mode of operation, insert the desired targeting parameters, and monitor related data throughout the mission. Essentially, the DEDA consists of a control panel to which electroluminescent displays and data entry pushbuttons are mounted and a logic enclosure that houses logic and input output circuits.

3.2.2.3. Abort Electronics Assembly. The AEA is a general-purpose, high-speed, 4,096-word digital computer that performs basic strapped-down guidance system calculations and the abort guidance and navigation steering calculations. The computer uses a fractional two's complement, parallel arithmetic section, and parallel data transfer. The AEA has three software computational sections: stabilization and alignment, navigation, and guidance.

The stabilization and alignment computational section computes stabilization and alignment on generation of mode signals by the DEDA. These mode signals determine the operation of the stabilization and alignment computational section in conjunction with the navigation and guidance computational sections.

The navigation computational section uses accelerometer inputs received from the ASA, via AEA input logic circuits, to calculate vehicle position and velocity in the inertial reference frame. The navigation computational section supplies total velocity, altitude, and altitude-rate data, and lateral velocity data in the vehicle reference frame, to the output logic circuits. Velocity data are routed to the DEDA, altitude-rate data are routed to the ALT RATE indicator, and lateral velocity data are routed to the X-pointer indicators. Velocity and position data are routed to the guidance computational section, for computing vehicle orbital parameters.

The guidance computational section provides trajectory computation and selection, steering computation, and midcourse-correction computation. This computational section receives data relating to the CSM state vector and the vehicle state vector from the LGC in other external source through the AGS input selector logic. Body-referenced steering errors are received from the stabilization and alignment computational section, for trajectory computation. The abort guidance problem consists of solving the equations of the selected guidance maneuver, including steering, attitude, and engine control computations. Outputs of the guidance computational section, through the output select logic circuits, include engine on and off signals to the CES, and velocity to be gained (selectable by DEDA readout).

Functionally, the AEA consists of a memory subassembly, central computer, an input-output subassembly, and a power subassembly.

3-2.3. CONTROL ELECTRONICS SECTION. (See figures 3.2.4 and 3.2.5.)

The CES comprises two attitude controller assemblies (ACA's), two thrust/translation controller assemblies (TTCA's), an attitude and translation control assembly (ATCA), a rate gyro assembly (RGA), descent engine control assembly (DECA), and three stabilization and control (S&C) control assemblies.

3-2.3.1. <u>Attitude Controller Assemblies.</u> The ACA's are right-hand pistol grip controllers, which the astronauts use to command changes in vehicle attitude. Each ACA is installed with its longitudinal axis approximately parallel to the X-axis. Each ACA supplies attitude rate commands proportional to the displacement of its handle, to the LGC and the ATCA; an out-of-detent discrete each time the handle is out of its neutral position; and a followup discrete to the AGS each time the controller is out of detent. A trigger-type push-to-talk switch on the pistol grip handle of the ACA is used for communication with the CSM and ground facilities.

As the astronaut uses his ACA, his hand movements are analogous to vehicle rotations. Clockwise or counterclockwise rotation of the controller commands yaw right or yaw left, respectively. Forward or aft movement of the controller commands vehicle pitch down or up, respectively. Left or right movement of the controller commands roll left or right, respectively.

The ACA's are also used in an incremental landing point designator (LPD) mode, which is available to the astronauts during the final approach phase. In this mode, the angular error between the designated landing site and the desired landing site is nulled by repetitive manipulation of an ACA. LPD signals from the ACA are directed to the LGC, which issues commands to move the designated landing site incrementally along the Y-axis and Z-axis.

3-2.3.2. Thrust/Translation Controller Assemblies. The TTCA's control LM translation in any axis; they are functionally integrated translation and thrust controllers. The astronauts use these assemblies to command vehicle translations by firing RCS thrusters and to throttle the descent engine between 10% and 92.5% thrust magnitude. The controllers are three-axis, T-handle, left-hand controllers, mounted with their longitudinal axis approximately 45° from a line parallel to the LM Z-axis (forward axis).

A lever on the right side of the TTCA enables the astronaut to select either of two control functions: (1) translation control in the Y-axis and Z-axis using the RCS thrusters and descent engine throttling to control X-axis translation and (2) translation control in all three axes using the RCS thrusters. Due to the TTCA mounting position, vehicle translations correspond to astronaut hand movements when operating the controller. Moving the T-handle to the left or right commands translation along the Y-axis. Moving the T-handle inward or outward commands translation along the Z-axis. Moving the T-handle upward or downward commands translation along the X-axis, using the RCS thrusters when the select lever is in the down position. When the lever is in the up position, upward or downward movement of the TTCA increases or decreases, respectively, the magnitude of descent engine thrust.

The TTCA is spring loaded to its neutral position in all axes when the lever is in the jets position. When the lever is in the throttle position, the Y- and Z-axis movements are spring loaded to the neutral position but the X-axis throttle commands will remain at the position set by the astronauts.

3-2.3.3. Attitude and Translation Control Assembly. The ATCA controls vehicle attitude and translation. In the primary guidance path, attitude and translation commands are generated by the LGC and applied directly to get drivers within the assembly. In the abort guidance path, the ATCA receives translation commands from a TTCA, rate-damping signals from the RGA, and attitude rate commands and pulse commands from the ACA. The ATCA combines attitude and translation commands in its logic network to select the proper thruster to be fired for the desired translation and/or rotation.

The ATCA routes the RCS thruster on and off commands from the LGC to the thrusters, in the primary control mode. During abort guidance control, the ATCA acts as a computer in determining which RCS thrusters are to be fired.

3-2.3.4. Rate Gyro Assembly. The RGA supplies the ATCA with damping signals to limit vehicle rotation rates and facilitates manual rate control during abort guidance control.

3-2.3.5. Descent Engine Control Assembly. The DECA processes engine-throttling commands from the astronauts (manual control) and the LGC (automatic control), gimbal commands for thrust vector control, preignition (arming) commands, and on and off commands to control descent engine operation.

The DECA accepts engine-on and engine-off commands from the S&C control assemblies, throttle commands from the LGC and the TTCA, and trim commands from the LGC or the ATCA. Demodulators, comparators, and relay logic circuits convert these inputs to the required descent engine commands. The DECA applies throttle and engine control commands to the descent engine and routes trim commands to the gimbal drive actuators.

3-2.3.6. S&C Control Assemblies. The S&C control assemblies are similar assemblies. They process, switch, and/or distribute the various signals associated with the GN&CS.

3-2.4. FUNCTIONAL DESCRIPTION.

The GN&CS comprises two functional loops, each of which is an independant guidance and control path. The primary guidance path contains elements necessary to perform all functions required to complete the lunar mission. If a failure occurs in this path the abort guidance path can be substituted.

3-2.4.1. Primary Guidance Path. (See figure 3-2.4.) The primary guidance path comprises the PGNS, CES, LR, RR, and the selected propulsion section required to perform the desired maneuvers. The CES routes flight control commands from the PGNS and applies them to the descent or ascent engine, and/or the appropriate thrusters.

The IMU which continuously measures attitude and acceleration, is the primary inertial sensing device of the vehicle. The LR senses slant range and velocity. The RR coherently tracks the CSM to derive LOS range, range rate, and angle rate. The LGC uses AOT star-sighting data to align the IMU. Using inputs from the LR, IMU, RR, TTCA's, and ACA's, the LGC solves guidance, navigation, steering, and stabilization equations necessary to initiate on and off commands for the descent and ascent engines, throttle commands and trim commands for the descent engine, and on and off commands for the thrusters.

Control of the vehicle when using the primary guidance path, ranges from fully automatic to manual. The primary guidance path operates in the automatic mode or the attitude hold mode. In the automatic mode, all navigation, guidance, stabilization, and control functions are controlled by the LGC. When the attitude hold mode is selected, the astronaut uses his ACA to bring the vehicle to a desired attitude. When the ACA is moved out of the detent position, proportional attitude-rate or minimum impulse commands are routed to the LGC. The LGC then calculates steering information and generates thruster commands that correspond to the mode of operation selected via the

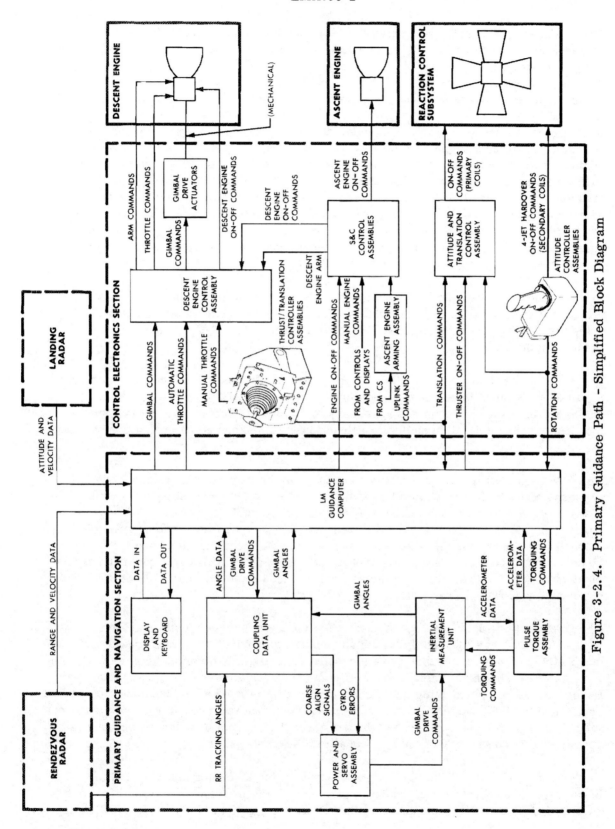

Figure 3-2.4. Primary Guidance Path – Simplified Block Diagram

DSKY. These commands are applied to the primary preamplifiers in the ATCA, which routes the commands to the proper thruster. When the astronaut releases the ACA, the LGC generates commands to hold this attitude. If the astronaut commands four-jet direct operation of the ACA by going to the hardover position, the ACA applies the command directly to the secondary solenoids of the corresponding thruster.

In the automatic mode, the LGC generates descent engine throttling commands, which are routed to the descent engine via the DECA. The astronaut can manually control descent engine throttling with his TTCA. The DECA sums the TTCA throttle commands with the LGC throttle commands and applies the resultant signal to the descent engine. The DECA also applies trim commands, generated by the LGC, to the GDA's to provide trim control of the descent engine. The LGC supplies on and off commands for the ascent and descent engines to the S&C control assemblies. The S&C control assemblies route the ascent engine on and off commands directly to the ascent engine, and the descent engine on and off commands to the descent engine via the DECA.

In the automatic mode, the LGC generates +X-axis translation commands to provide ullage. In the manual mode, manual translation commands are generated by the astronaut, using his TTCA. These commands are routed, through the LGC, to the ATCA and on to the proper thruster.

3-2.4.2. Abort Guidance Path. (See figure 3-2.5.) The abort guidance path comprises the AGS, CES, and the selected propulsion section. The AGS performs all inertial navigation and guidance functions necessary to effect a safe orbit or rendezvous with the CSM. The stabilization and control functions are performed by analog computation techniques, in the CES.

The AGS uses a strapped-down inertial sensor, rather than the stabilized, gimbaled sensor used in the IMU. The ASA is a strapped-down inertial sensor package that measures attitude and acceleration with respect to the vehicle body axes. The ASA-sensed attitude is supplied to the AEA, which is a high-speed, general-purpose digital computer that performs the basic strapped-down system computations and the abort guidance and navigation steering control calculations. The DEDA is a general-purpose input-output device through which the astronaut manually enters data into the AEA and commands various data readouts.

The CES functions as an analog autopilot when the abort guidance path is selected. It uses inputs from the AGS and from the astronauts to provide the following: on, off, and TTCA throttling commands for the descent engine; gimbal commands for the GDA's to control descent engine trim; on and off commands for the ascent engine; sequencer logic to ensure proper arming and staging before engine startup and shutdown; on and off commands for the thrusters for translation and stabilization, and for various maneuvers; jet-select logic to select the proper thrusters for the various maneuvers; and modes of vehicle control, ranging from fully automatic to manual.

The astronaut uses the TTCA to control descent engine throttling and translation maneuvers. The throttle commands, engine on and off commands from the S&C control assemblies, and trim commands from the ATCA are applied to DECA. The DECA applies

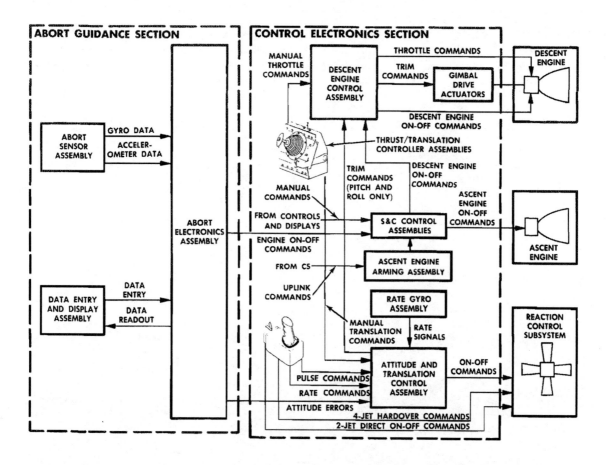

Figure 3-2.5. Abort Guidance Path - Simplified Block Diagram

the throttle commands to the descent engine, the engine on and off commands to the descent
engine latching device, and the trim commands to the GDA's. The S&C control assemblies
receive engine on and off commands for the descent and ascent engines from the AEA. As
in the primary guidance path, the S&C control assemblies route descent engine on and off
commands to the DECA and apply ascent engine on and off commands directly to the ascent
engine.

The abort guidance path operates in the automatic mode or the attitude hold mode. In the
automatic mode, navigation and guidance functions are controlled by the AGS; stabilization
and control functions, by the CES. In the attitude hold mode, the astronaut uses his ACA to
control vehicle attitude. The ACA generates attitude-rate, pulse, direct, and hardover
commands. The attitude-rate and pulse commands, AEA error signals, RGA rate-damping
signals, and TTCA translation commands are applied to the ATCA. The ATCA processes
these inputs to generate thruster on and off commands.

In the attitude hold mode, pulse and direct submodes are available for each axis. The
pulse submode is an open-loop attitude control mode in which the ACA is used to make
small attitude changes in the selected axis. The direct submode is an open-loop attitude
control mode in which pairs of thrusters are directly controlled by the ACA. The

astronaut can also control vehicle attitude in any axis by moving the ACA to the hardover position. In addition, the astronaut can override translation control in the +X-axis with a +X-axis translation pushbutton. Pressing the pushbutton fires all four +X-axis thrusters.

3-3. RADAR SUBSYSTEM.

During the landing phase and subsequent rendezvous phase, the LM uses radar navigational techniques to determine distance and velocity. Each phase uses a radar designed specifically for that phase (landing radar, rendezvous radar). Both radars inform the
astronaut and the computer concerning position and velocity relative to acquired target.
During lunar landing, the target is the surface of the moon; during rendezvous, the target is
the Command Module.

3-3.1. LANDING RADAR.

The landing radar, located in the descent stage, provides altitude and velocity data during
lunar descent. The primary guidance and navigation section calculates control signals for
descent rate, hovering, and safe landing. For the LM, altitude data begins at approximately
38,000 feet above the lunar surface; velocity data, at approximately 25,000 feet. (Refer
to table 3-3.1.) The landing radar uses four microwave beams: three, to measure velocity
by Doppler shift continuous wave; one, to measure altitude by continuous-wave frequency
modulation. (See figure 3-3.1.)

The landing radar senses the velocity and altitude of the vehicle relative to the lunar surface
by means of a three-beam Doppler velocity sensor and a single-beam radar altimeter.
Velocity and range data are made available to the LM guidance computer as 15-bit binary
words; forward and lateral velocity data, to the displays as d-c analog voltages; and range
and range rate data, to the displays as pulse-repetition frequencies.

The landing radar consists of an antenna assembly and an electronics assembly. The antenna assembly forms, directs, transmits, and receives the four microwave beams. Two
interlaced phased arrays transmit the velocity and altimeter-beam energy. Four broadside
arrays receive the reflected energy of the three velocity beams and the altimeter beam.
The electronics assembly processes the Doppler and continuous-wave FM returns, which
provide the velocity and slant range data for the LM guidance computer and the displays.

The antenna assembly transmits velocity beams (10.51 GHz) and an altimeter beam (9.58
GHz) to the lunar surface.

When the electronics assembly is receiving and processing the returned microwave beams,
data-good signals are sent to the LM guidance computer. When the electronics assembly is
not operating properly, data-no-good signals are sent to the pulse-code-modulation and
timing electronics assembly of the Instrumentation Subsystem, for telemetry.

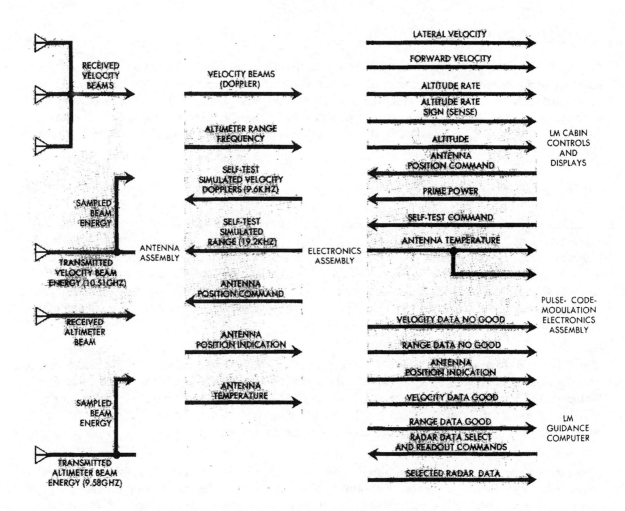

Figure 3-3.1.   Landing Radar-Signal Flow

Using controls and indicators, the astronauts can monitor vehicle velocity, altitude, and radar-transmitter power and temperatures; apply power to energize the radar; initiate self-test; and place the antenna in the descent or hover position.   Self-test permits operational checks of the radar without radar returns from external sources.   An antenna temperature control circuit, energized at earth launch, protects antenna components against the low temperatures of space environment while the radar is not operating.

The radar is first turned on and self-tested during vehicle checkout before separation from the CSM.   The self-test circuits apply simulated Doppler signals to radar velocity sensors, and simulated lunar range signals to an altimeter sensor.   The radar is self-tested again immediately before powered descent, approximately 70,000 feet above the lunar surface. The radar operates from approximately 50,000 feet until lunar touchdown.

Altitude (derived from slant range) and forward and lateral velocities are available to the LM guidance computer and cabin indicators. Slant range data are continuously updated to provide true altitude above the lunar surface.

At approximately 200 feet above the lunar surface, the vehicle pitches to orient its X-axis perpendicular to the surface; all velocity vectors are near zero. Final visual selection of the landing site is followed by touchdown under automatic or manual control. During this phase, the astronauts monitor altitude and velocity data from the radar.

The landing radar antenna has a descent position and a hover position. In the descent position, the antenna boresight angle is 24° from the LM X-axis. In the hover position, the antenna boresight is parallel to the X-axis and perpendicular to the Z-axis. Antenna position is selected by the astronaut during manual operation and by the LM guidance computer during automatic operation. During automatic operation, the LM guidance computer commands the antenna to the hover position 8,000 to 9,000 feet above the lunar surface.

3-3.1.1. Antenna Assembly. The assembly comprises four microwave mixers, four dual audio-frequency preamplifiers, two microwave transmitters, a frequency modulator, and and antenna pedestal tilt mechanism.

The antenna consists of six planar arrays: two, for transmission; four, for reception. They are mounted on the tilt mechanism, beneath the descent stage, and may be placed in one of two fixed positions.

3-3.1.2. Electronics Assembly. The electronics assembly comprises frequency trackers (one for each velocity beam), a range frequency tracker, velocity converter and computer, range computer, signal data converter, and data-good/no-good logic circuit.

3-3.2. RENDEZVOUS RADAR.

The rendezvous radar, operated in conjunction with a CSM transponder, acquires and tracks the CSM before and during rendezvous and docking. The radar, located in the ascent stage,

Table 3-3.1. Landing Radar Data

Parameter	Current Design
Total travel angle	24°
Maximum travel limits	+24° to 0°
Antenna angles	+24°, 0°
Velocity data good	25,000 ft
Altitude data good	38,000 ft
Zero Doppler dropouts	350 to 135 ft

tracks the CSM during the descent phase of the mission to supply tracking data for any required abort maneuver and during the ascent phase to supply data for rendezvous and docking. When the radar tracks the CSM, continuous measurements of range, range rate, angle, and angle rate (with respect to the LM) are provided simultaneously to the primary guidance and navigation section and to LM cabin displays. This allows rendezvous to be performed automatically under computer control, or manually by the astronauts. During the rendezvous phase, rendezvous radar performance is evaluated by comparing radar range and range rate tracking values with MSFN tracking values.

The CSM transponder receives an X-band three-tone phase-modulated, continuous-wave signal from the rendezvous radar, offsets the signal by a specified amount, and then transmits a phase-coherent carrier frequency for acquisition by the radar. This return signal makes the CSM appear as the only object in the radar field of view. The transponder provides the long range (400 nm) required for the mission.

The transponder and the radar use solid-state varactor frequency-multiplier chains as transmitters, to provide high reliability. The radar antenna is space stabilized to negate the effect of LM vehicle motion on the line-of-sight angle. The gyros used for this purpose are rate-integrating types; in the maual mode they also supply accurate line-of-sight, angle-rate data for the astronauts. Range rate is determined by measuring the two-way Doppler frequency shift on the signal received from the transponder. Range is determined by measuring the time delay between the received and the transmitted three-tone phase-modulated waveform.

The rendezvous radar has an antenna assembly and an electronics assembly. The antenna assembly automatically tracks the transponder signal after the electronics assembly acquires the transponder carrier frequency. The return signal from the transponder is received by a four-port feedhorn. The feedhorn, arranged in a simultaneous lobing configuration, is located at the focus of a Cassegrainian antenna. If the transponder is directly in line with the antenna boresight, the transponder signal energy is equally distributed to each port of the feedhorn. If the transponder is not directly in line, the signal energy is unequally distributed among the four ports.

The signal passes through a polarization diplexer to a comparator, which processes the signal to develop sum and difference signals. The sum signal represents the sum of energy received by all feedhorn ports (A + B + C + D). The difference signals, representing the difference in energy received by the feedhorn ports, are processed along two channels: a shaft-difference channel and a trunnion-difference channel. The shaft-difference signal represents the vectoral sum of the energy received by adjacent ports (A + D) - (B + C) of the feedhorn. The trunnion-difference signal represents the vectoral sum of the energy received by adjacent ports (A + B) - (C + D). The comparator outputs are heterodyned with the transmitter frequency to obtain three intermediate-frequency signals. After further processing, these signals provide unambiguous range, range rate, and direction of the CSM. This information is fed to the LM guidance computer and to cabin displays. (See figure 3-3.2.)

The rendezvous radar operates in three modes: automatic tracking, slew (manual), or LM guidance computer control.

The automatic tracking mode enables the radar to track the CSM automatically after it has been acquired; tracking is independent of LM guidance computer control. When this mode is selected, tracking is maintained by comparing the received signals from the shaft and trunnion channels with the sum channel signal. The resultant error signals drive the antenna, thus maintaining track.

The slew mode enables an astronaut to position the antenna manually to acquire the CSM.

In the LM guidance computer control mode, the computer automatically controls antenna positioning, initiates automatic tracking once the CSM is acquired, and controls change in antenna orientation. The primary guidance and navigation section, which transmits computer-derived commands to position the radar antenna, provides automatic control of radar search and acquisition.

3-3.2.1. Antenna Assembly. The main portion of the rendezvous radar antenna is a 24-inch parabolic reflector. A 4.65-inch hyperbolic subreflector is supported by four converging struts. Before the radar is used, the antenna is manually released from its stowed position. The antenna pedestal and the base of the antenna assembly are mounted on the external structural members of the vehicle. The antenna pedestal includes rotating assemblies that contain radar components. The rotating assemblies are balanced about a shaft axis and a trunnion axis. The trunnion axis is perpendicular to, and intersects, the shaft axis. The antenna reflectors and the microwave and RF electronics components are assembled at the top of the trunnion axis. This assembly is counterbalanced by the trunnion-axis rotating components (gyroscopes, resolvers, and drive motors) mounted below the shaft axis. Both

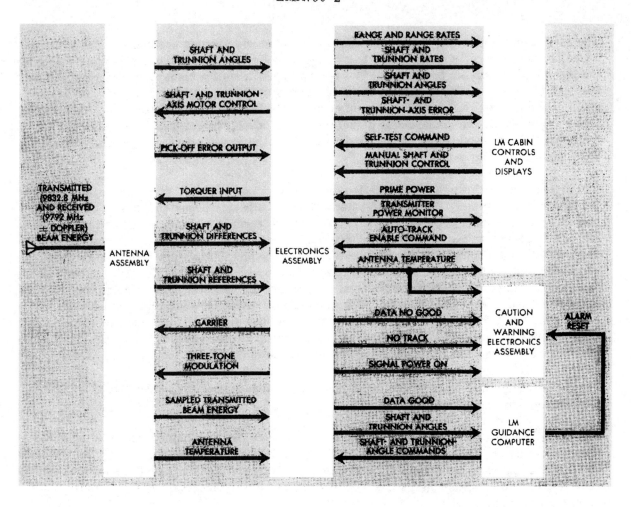

Figure 3-3.2. Rendezvous Radar-Signal Flow

groups of components, mounted opposite each other on the trunnion axis, revolve about the shaft axis. This balanced arrangement requires less driving torque and reduces the overall antenna weight. The microwave, radiating, and gimbaling components, and other internally mounted components, have low-frequency flexible cables that connect the outboard antenna components to the inboard electronics assembly.

3-3.2.2. <u>Electronics Assembly</u>. The electronics assembly comprises a receiver, frequency synthesizer, frequency tracker, range tracker, servo electronics, a signal data converter, self-test circuitry, and a power supply. The assembly furnishes crystal-controlled signals, which drive the antenna assembly transmitter; provides a reference for receiving and processing the return signal; and supplies signals for antenna positioning.

3-4. <u>MAIN PROPULSION SUBSYSTEM.</u>

The Main Propulsion Subsystem (MPS) consists of two separate, complete, and independent propulsion sections: the descent propulsion section and the ascent propulsion section. (See figure 3-4.1.) The descent propulsion section provides propulsion for the LM from the time it separates from the CSM until it lands on the lunar surface. The ascent propulsion section lifts the ascent stage off the lunar surface and boosts it into orbit. Both propulsion sections operate in conjunction with the Reaction Control Subsystem (RCS), which provides propulsion used mainly for precise attitude and translation maneuvers. If a mission abort becomes necessary during the descent trajectory, the ascent or descent engine can be used to return to a rendezvous orbit with the CSM. The choice of engines depends on the cause for abort, the amount of propellant remaining in the descent stage, and the length of time that the descent engine had been firing.

Each propulsion section consists of a liquid-propellant, pressure-fed rocket engine and propellant storage, pressurization, and feed components. For reliability, many vital components in each section are redundant. In both propulsion sections, pressurized helium forces the hypergolic propellants from the tanks to the engine injector. Both engine assemblies have control valves and trim orifices that start and stop a metered propellant flow to the combustion chamber upon command, an injector that determines the spray pattern of the propellants as they enter the combustion chamber, and a combustion chamber, where the propellants meet and ignite. The gases produced by combustion pass through a throat area into the engine nozzle, where they expand at an extremely high velocity before being ejected. The momentum of the exhaust gases produces the reactive force that propels the vehicle.

The more complicated tasks required of the descent propulsion section dictate that the descent section be the larger and more sophisticated of the two propulsion sections. In the modified vehicle the descent propellant tanks hold 18,697 pounds of usable propellants (compared to 17,489 pounds in the LM), and amount that is more than three times that of the ascent propulsion section. The descent engine is almost twice as large as the ascent engine, produces more thrust (almost 10,000 pounds at full throttle), is throttleable for thrust control, and is gimbaled for thrust vector control. The ascent engine, which cannot be tilted, delivers a fixed thrust of 3,500 pounds, sufficient to launch the ascent stage from the lunar surface and place it into a predetermined orbit.

3-4.1. PROPELLANTS.

The ascent and descent propulsion sections, as well as the RCS, use identical fuel/oxidizer combinations. In the ascent and descent propulsion sections, the injection ratio of oxidizer to fuel is approximately 1.6 to 1, by weight.

The fuel is a blend of hydrazine (N_2H_4) and unsymmetrical dimethylhydrazine (UDMH), commercially known as Aerozine 50. The proportions, by weight, are approximately 50% hydrazine, and 50% dimethylhydrazine.

The oxidizer is nitrogen tetroxide (N_2O_4). It has a minimum purity of 99.5% and a maximum water content of 0.1%.

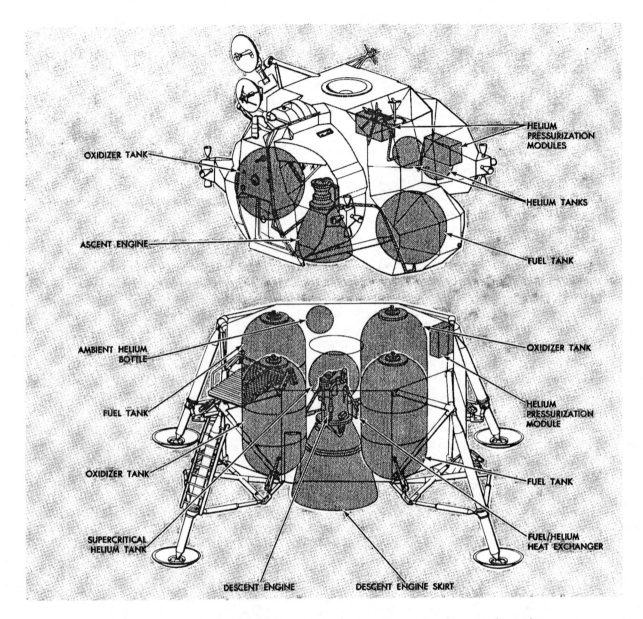

Figure 3-4.1. Main Propulsion Subsystem - Major Equipment Location

3-4.2. MAIN PROPULSION SUBSYSTEM OPERATION.

The MPS is operated by the Guidance, Navigation, and Control Subsystem (GN&CS), which issues automatic (and processes manually initiated) on and off commands to the descent or ascent engine. The GN&CS also furnishes gimbal-drive and thrust-level commands to the descent propulsion section.

Before starting either engine, the propellants must be settled to the bottom of the tanks. Under weightless conditions, this requires an ullage maneuver; that is, the vehicle must be moved in the +X, or upward, direction. To perform this maneuver, the downward-firing thrusters of the RCS are operated. The duration of this maneuver increases for each engine start because more time is required to settle the propellants as the tanks become emptier.

3-4.3. DESCENT PROPULSION SECTION.

The descent propulsion section consists of the helium pressurization components; two fuel and two oxidizer tanks with associated feed components; and a pressure-fed, ablative, throttleable rocket engine. The engine can be shut down and restarted as required by the mission. At the full-throttle position, the engine develops a nominal thrust of 9,870 pounds; it can also be operated within a range of 1,050 to 6,300 pounds of thrust.

3-4.3.1. Descent Engine Operation and Control. After initial pressurization of the descent propulsion section, the descent engine start requires two separate and distinct operations: arming and firing. Engine arming is performed by the astronauts; engine firing can be initiated manually by the astronauts, or automatically by the LM guidance computer. When the astronauts set a switch to arm the descent engine, power is simultaneously routed to open the actuator isolation valves in the descent engine, enable the instrumentation circuits in the descent propulsion section, and issue a command to the throttling controls to start the descent engine at the required 10% thrust level. The LM guidance computer and the abort guidance section receive an engine-armed status signal. This signal enables an automatic engine-on program in the GN&CS, resulting in a descent engine start. A manual start is accomplished when the Commander pushes his engine-start pushbutton. (Either astronaut can stop the engine because separate engine-stop pushbuttons are provided at both flight stations.)

The normal profile for all descent engine starts must be at 10% throttle setting because the thrust vector at engine start may not be directed through the vehicle center of gravity. A low-thrust start (10%) permits corrective gimbaling. If the engine is started at high thrust, RCS propellants must be used to stabilize the vehicle.

The astronauts can, with panel controls, select automatic or manual throttle control modes and Commander or LM Pilot thrust/translation controller authority, and can override automatic engine operation. Redundant circuits, under astronaut control, ensure descent engine operation if prime control circuits fail.

Signals from the GN&CS automatically control descent engine gimbal trim a maximum of 6° from the center position in the Y-axis and Z-axis to compensate for center-of-gravity offsets during descent engine firing. The gimbal ring is located at the plane of the descent engine combustion chamber throat. Two gimbal drive actuators tilt the descent engine in the gimbal ring. The actuators can extend or retract 2 inches from the midposition. One actuator controls the pitch gimbal; the other the roll gimbal.

3-4.3.2. Pressurization Section. (See figure 3-4.2.) Before earth launch, the LM propellant tanks are only partly pressurized (less than 230 psia), so that the tanks will be maintained within a safe pressure level under the temperature changes experienced between

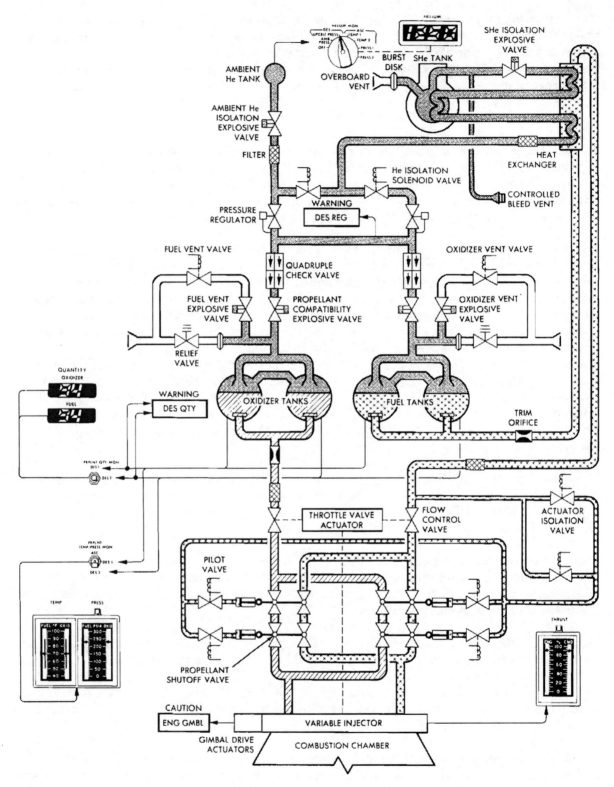

Figure 3-4.2.   Descent Propulsion Section - Flow Diagram

filling and launch. At initial engine start, the ullage space in each propellant tank requires additional pressurization. This initial pressurization is accomplished with a relatively small amount of helium stored at ambient temperature and at an intermediate pressure.

Supercritical helium is stored at a density approximately eight times that of ambient helium. Because heat transfer from the outside to the inside of the cryogenic storage vessel causes a gradual increase in pressure (approximately 10 psi per hour maximum), the initial loading pressure is planned so that the supercritical helium will be maintained within a safe pressure/time envelope throughout the mission.

The supercritical helium tank has a burst disk assembly and an internal helium/helium heat exchanger. The burst disk assembly prevents hazardous overpressurization within the vessel. It consists of two burst disks in series, with a normally open, low-pressure vent valve between the disks. The burst disks rupture at a pressure between 1,881 and 1,967 psid to vent the entire supercritical helium supply overboard. A thrust neutralizer at the outlet of the downstream burst disk diverts the escaping gas into opposite directions to prevent unidirectional thrust generation. The vent valve prevents low-pressure buildup between the burst disks if the upstream burst disk leaks slightly. In addition to this venting arrangement, the modified LM has a controlled bleed vent in parallel. This bleed vent - a series of stacked perforated plates - permits a small controlled amount of supercritical helium to be bled, extending the supercritical helium tank maximum standby time from 131 hours to 190 hours.

To open the pressurization path to the propellant tanks, an explosive helium isolation valve and two propellant compatibility values must be fired. The helium isolation valve is automatically fired 1.3 seconds after the descent engine start command is issued. The time delay prevents the supercritical helium from entering the fuel/helium heat exchanger until propellant flow is established so that the fuel cannot freeze in the heat exchanger. The supercritical helium enters the two-pass fuel/helium heat exchanger where it is slightly warmed by the fuel. The helium then flows back into a heat exchanger in the supercritical helium tank where it increases the temperature of the supercritical helium in the tank. Finally, the helium flows through the second loop of the fuel/helium heat exchanger where it is heated to operational temperature.

After flowing through a filter, the helium enters a pressure regulator system which reduces the helium pressure to approximately 245 psi. This system consists of two parallel, redundant regulators. The regulated helium then enters parallel paths, which lead through quadruple check valves into the propellant tanks. The quadruple check valves, consisting of four valves in a series-parallel arrangement, permit flow in one direction only. This protects upstream components against corrosive propellant vapors and prevents hypergolic action due to backflow from the propellant tanks.

A relief valve, which opens at approximately 268 psia, protects each propellant tank against overpressurization. A thrust neutralizer prevents the gas from generating unidirectional thrust. Each relief valve is paralleled by two series-connected vent valves. After landing, the astronauts open the vent valves to relieve pressure buildup in the tanks

3-4.3.3 Propellant Feed Section. The descent section propellant supply is contained in four cylindrical, spherical-ended titanium tanks of identical size and construction. Two tanks contain fuel; the other two, oxidizer. In the LM, each pair of tanks containing like propellants is interconnected at the top and bottom to ensure even distribution of propellant and pressurizing helium. For the modified LM, the double crossfeed arrangement at the tank outlets is eliminated by removal of the propellant balance lines. A diffuser at the helium inlet port of each tank uniformly distributes the pressurizing helium into the tank. An antivortex device in the form of a series of vanes, at each tank outlet, prevents the propellant from swirling into the outlet port, thus precluding inadvertent helium ingestion into the engine. Each tank outlet also has a propellant retention device (negative-g can) that permits unrestricted propellant flow from the tank under normal pressurization, but blocks reverse propellant flow under zero-g or negative-g conditions.

Pressurized helium, acting on the surface of the propellant, forces the fuel and oxidizer into the delivery lines. The oxidizer is piped directly to the engine assembly; the fuel circulates through the fuel/helium heat exchanger (where it warms the supercritical helium) before it is routed to the engine assembly. Each delivery line contains a trim orifice and a filter. The trim orifices provide engine inlet pressure of approximately 220 psia at full throttle position. The filters prevent debris from contaminating downstream components.

3-4.3.4 Propellant Quantity Gaging System. The propellant quantity gaging system enables the astronauts to monitor the quantity of propellants remaining in the four descent tanks. It is in operation during the final powered descent phase, from start of the braking maneuver (10 seconds after engine turn-on) until lunar touchdown. The system consists of four quantity-sensing probes with low-level sensors, a control unit, two quantity indicators, a switch that permits the astronauts to select a set of tanks (one fuel and one oxidizer) to be monitored, and a low-level warning light. The low-level sensors provide a discrete signal to cause the warning light to go on when the propellant level in any descent tank is down to 9.4 inches, an amount sufficient for 2 minutes of engine burn at hover thrust (approximately 25%).

3-4.3.5 Engine Assembly. (See figures 3-4.3 and 3-4.4.) The descent engine is mounted in the center compartment of the descent stage cruciform. Fuel and oxidizer entering the engine assembly are routed through flow control valves to the propellant shutoff valves. A total of eight propellant shutoff valves are used; they are arranged in series-parallel redundancy, four in the fuel line and four in the oxidizer line. The series redundancy ensures engine shutoff, should one valve fail to close. The parallel redundancy ensures engine start, should one valve fail to open.

The propellant shutoff valves are actuator operated. The actuation line branches off the main fuel line at the engine inlet and routes fuel through actuator isolation valves and pilot valves into four hydraulically operated actuators. The actuator pistons are connected to rack-and-pinion linkages that rotate the balls of the shutoff valves 90° to the open position. The actuator isolation valves open when the astronauts arm the descent engine. When an engine-on command is initiated, the four pilot valves open simultaneously.

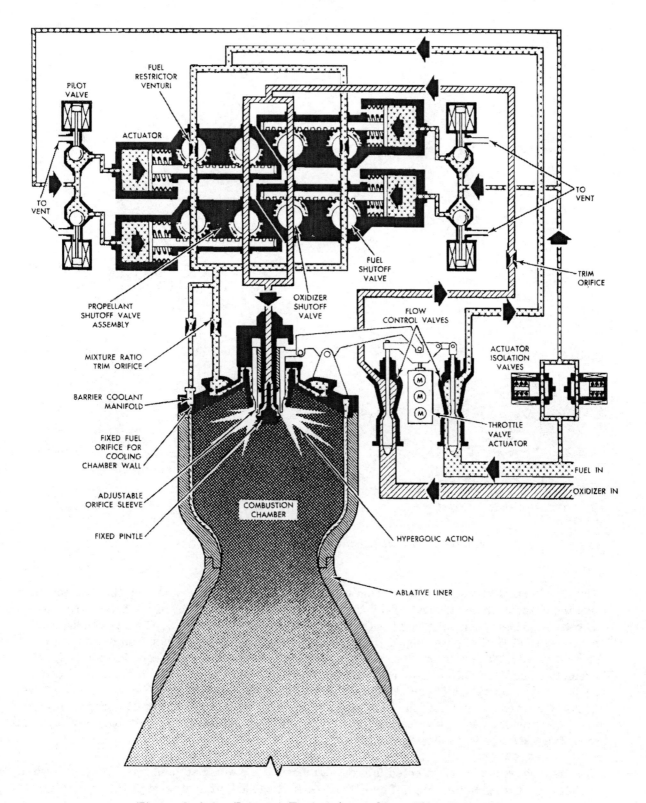

Figure 3-4.3.   Descent Engine Assembly – Flow Diagram

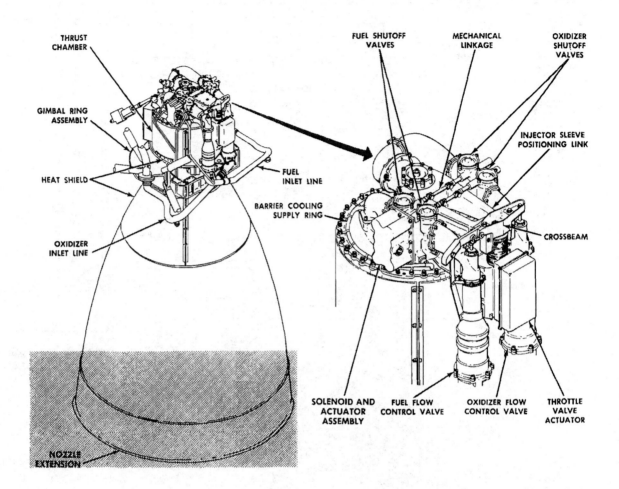

Figure 3-4.4. Descent Engine and Head End Assembly

The flow control valves, in conjunction with the adjustable orifice sleeve in the injector, control the descent engine thrust. They are adjusted simultaneously by a mechanical linkage. Throttle setting is controlled by the throttle valve actuator, which positions the linkage in response to electrical input signals, changing the position of the movable pintles in the flow control valves. The axial movement of the pintles decreases or increases the pintle flow areas to control propellant flow rate and thrust. In the throttling range between 60% and 92.5% thrust, operation of cavitating venturis of the flow control valves becomes uppredictable any may cause an improper fuel-oxidizer mixture ratio. Because an improper mixture ratio will result in excessive engine erosion and early combustion chamber burn-through, the throttling range between 60% and 92.5% thrust is not used.

The fuel and oxidizer are injected into the combustion chamber at velocities and angles compatible with variations in weight flow. Some fuel is tapped off upstream of the injector and is routed through a trim orifice into the barrier coolant manifold. From there, it is sprayed against the combustion chamber wall, maintaining the wall at an acceptable temperature.

The combustion chamber consists of an ablative-cooled chamber section, nozzle throat, and nozzle divergent section. The ablative sections are enclosed in a continuous titanium shell and jacketed in a thermal blanket. The nozzle extension is a radiation-cooled, crushable skirt, it can collapse a distance of 28 inches on lunar impact so as not to affect the stability of the vehicle. The nozzle extension is made of columbium coated with aluminide. It is attached to the combusion chamber case at a nozzle area ratio of 16 to 1 and extends to an exit area ratio of 47.4 to 1. For the modified LM, a permanently attached 10-inch-long extension is being developed. The increased length provides higher engine efficiency (Isp increased by approximately 2.5 seconds at full thrust.)

3-4.4. ASCENT PROPULSION SECTION

The ascent propulsion section consists of a constant-thrust, pressure-fed rocket engine, one fuel and one oxidizer tank, two helium tanks, and associated propellant feed and helium pressurization components. The engine develops 3,500 pounds of thrust in a vacuum, it can be shut down and restarted, as required by the mission.

3-4.4.1. <u>Ascent Engine Operation and Control</u>. The ascent engine, like the descent engine, requires manual arming before it can be fired. When the astronauts arm the ascent engine, a shutoff command is sent to the descent engine. Then, enabling signals are sent to the ascent engine control circuitry to permit a manual or computer-initiated ascent engine start. Shortly before initial ascent engine use, the astronauts fire explosive valves to pressurize the ascent propulsion section

For manual engine on and off commands, the astronauts push the same start and stop pushbuttons used for the descent engine. For automatic commands, the stabilization and control assemblies in the GN&CS provide sequential control of vehicle staging and ascent engine on and off commands. The initial ascent engine firing - whether for normal lift-off from the lunar surface or in-flight abort - is a fire-in-the-hole (FITH) operation; that is, the engine fires while the ascent and descent stages are still mated although no longer mechanically secured to each other. If, during the descent trajectory, an abort situation necessitates using the ascent engine to return to the CSM, the astronauts initiate an abort stage sequence. This results in an immediate descent engine shutdown followed by a time delay to ensure that the engine has stopped thrusting before staging occurs. The next command automatically pressurizes the ascent propellant tanks, after which the staging command is issued. This results in severing of hardware that secures the ascent stage to the descent stage and the interconnecting cables. The ascent engine fire command completes the abort stage sequence.

3-4.4.2. <u>Pressurization Section</u>. (See figure 3-4.5.) Before initial ascent engine start, the propellant tanks must be fully pressurized with gaseous helium. This helium is stored in two identical tanks at a nominal pressure of 3,050 psia at a temperature of +70° F. To open the helium paths to the propellant tanks, the astronauts normally fire six explosive valves simultaneously: two helium isolation valves and four propellant compatibility valves (two connected in parallel for redundancy in each pressurization path). Before firing the

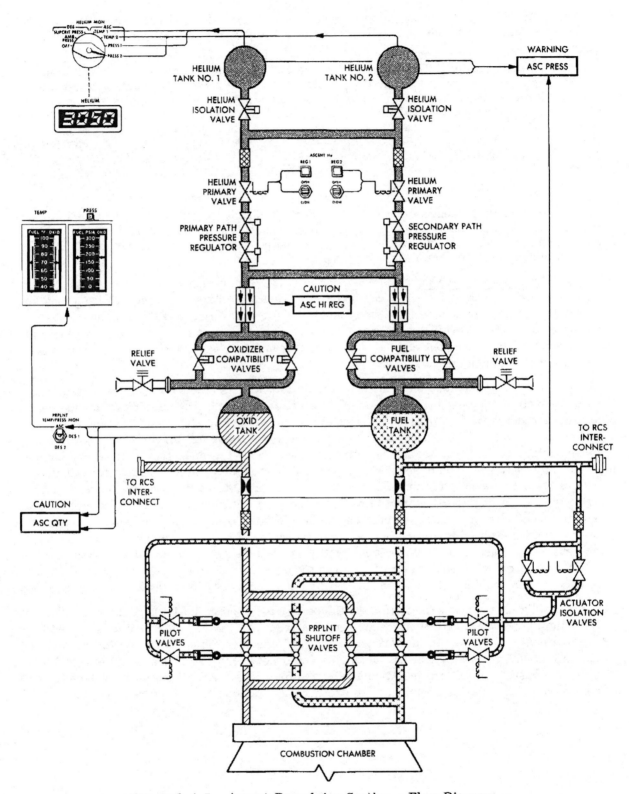

Figure 3-4.5. Ascent Propulsion Section - Flow Diagram

explosive valves, the astronauts check the pressure in each helium tank. If one tank provides an unusually low reading (indicating leakage), they can exclude the appropriate helium isolation explosive valve from the fire command.

The helium flows through the primary and secondary pressure regulating paths, each containing a filter, a normally open solenoid valve and two series-connected pressure regulators. Two downstream regulators are set to a slightly higher output pressure than the upstream regulators; the regulator pair in the primary flow path produces a slightly higher output than the pair in the secondary (redundant) flow path. This arrangement causes lockup of the regulators in the redundant flow path after the propellant tanks are pressurized, while the upstream regulator in the primary flow path maintains the propellant tanks at their normal pressure of 184 psia.

The regulated helium is routed into two flow paths: one path leads to the oxidizer tank; the other, to the fuel tank. A quadruple check valve assembly, a series-parallel arrangement in each path, isolates the upstream components from corrosive propellant vapors. The check valves also safeguard against possible hypergolic action in the common manifold, resulting from mixing of propellants or fumes flowing back from the propellant tanks. Each helium path contains a burst disk and relief valve assembly to protect the propellant tanks. This assembly vents pressure in excess of approximately 226 psi and reseals the flow path after overpressurization is relieved.

3-4.4.3. Propellant Feed Section. The ascent section propellant supply is contained in two spherical titanium tanks of identical size and construction. One tank contains fuel; the other, oxidizer. A helium diffuser at the inlet port of each tank uniformly distributes the pressurizing helium into the tank. An antivortex device (a cruciform at each tank outlet) prevents the propellant from swirling into the outlet port, precluding helium ingestion into the engine. Each tank outlet also has a propellant-retention device that permits unrestricted propellant flow from the tank under normal pressurization, but blocks reverse propellant flow under zero-g or negative-g conditions. Low-level sensors cause a caution light to go on when the propellant remaining in either tank is sufficient for approximately 10 seconds of burn time (43 pounds of fuel, 69 pounds of oxidizer). Transducers enable the astronauts to monitor propellant temperature and ullage pressure.

The outflow from each propellant tank divides into two paths. The primary path routes each propellant through a trim orifice and a filter to the propellant shutoff valves in the engine assembly. The trim orifice provides an engine inlet pressure of 165 psia for proper propellant use. The secondary path connects the ascent propellant supply to the RCS. A line branches off the RCS interconnect fuel path and leads through two parallel actuator isolation solenoid valves to the engine pilot valves.

3-4.4.4. Engine Assembly. (See figures 3-4.6 and 3-4.7.) The ascent engine is installed in the midsection of the ascent stage; it is tilted so that its centerline is 1.5° from the X-axis, in the +Z-direction. Fuel and oxidizer entering the engine assembly are routed through filters, propellant shutoff valves, and trim orifices, to the injector. A separate fuel path leads from the actuator isolation valves to the pilot valves. The fuel in this line enters the actuators, which open the propellant shutoff valves.

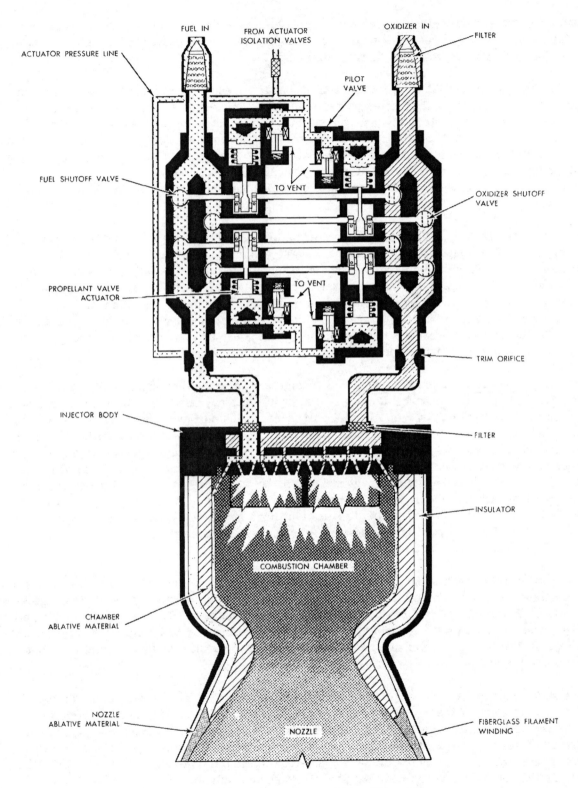

Figure 3-4.6. Ascent Engine Assembly - Flow Diagram

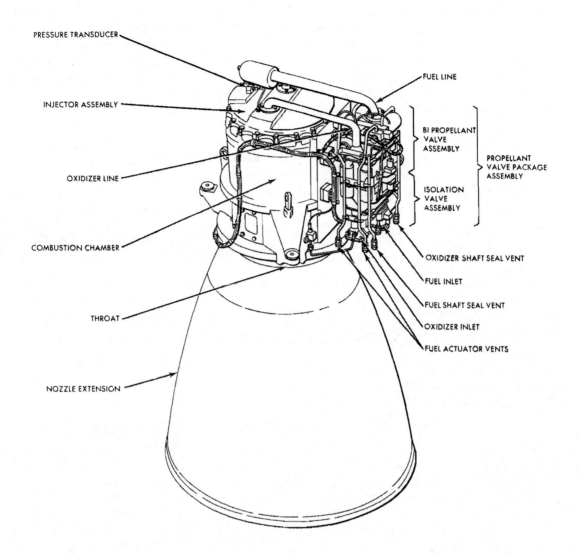

Figure 3-4.7. Ascent Engine Assembly

The valve package that controls propellant flow into the injector consists of eight propellant shutoff valves (four fuel and four oxidizer in a series-parallel arrangement to provide re-dundant flow paths and shutoff capability) and four solenoid-operated pilot valve and actuator assemblies. One fuel shutoff valve and one oxidizer shutoff valve are operated by a common shaft, which is connected to its respective pilot valve and actuator assembly. Shaft seals and vented cavities prevent the propellants from coming into contact with each other. Sep-arate overboard vents drain the fuel and oxidizer that leaks past the valve seals. The eight shutoff valves open and close simultaneously to permit or terminate propellant flow to the injector. The four nonlatching, solenoid-operated pilot valves control the actuator fluid.

When an engine-start command is received, the two actuator isolation valves and the four pilot valves open simultaneously and fuel flows into the actuator chambers. Hydraulic pressure extends the actuator pistons, cranking the propellant shutoff ball valves 90° to the fully open position. The propellants flow through a final set of orificies that trim the pressure differentials of the fuel and oxidizer to determine the mixture ratio of the propellants as they reach the injector. The injector establishes an oxidizer lead of approximately 70 milliseconds, precluding the possibility of a rough engine start.

The combustion chamber wall is cooled by spraying fuel against it through canted orifices spaced around the perimeter of the injector. The combustion chamber ablative material extends from the injector to an expansion ratio of 4.6. The chamber insulator, between the ablative material and the engine case, maintains the chamber skin temperature within design requirements. The ablative material of the nozzle extension extends from the expansion ratio of 4.6 to 45.6 (exit plane) and provides ablative cooling in this region. For the modified LM, a recontoured ablative nozzle is being developed. This will result in weight saving and improved performance.

At engine shutdown, the actuator isolation valves are closed, preventing additional fuel from reaching the pilot valves. Simultaneously, the pilot valve solenoids are deenergized, opening the actuator ports to the overboard vents so that residual fuel in the actuators is vented into space. With the actuation fluid pressure removed the actuator pistons are forced back by spring pressure, cranking the propellant shutoff valves to the closed position.

3-5. REACTION CONTROL SUBSYSTEM.

The Reaction Control Subsystem (RCS) provides thrust impulses that stabilize the LM during the descent and ascent trajectory and controls attitude and translation - movement of the vehicle about and along its three axes - during hover, landing, and rendezvous and docking maneuvers. The RCS also provides the thrust required to separate the LM from the CSM and the +X-axis acceleration (ullage maneuver) required to settle Main Propulsion Subsystem (MPS) propellants before a descent or ascent engine start. The RCS accomplishes its task during coasting periods or while the descent or ascent engine is firing; it operates in response to automatic control commands from the Guidance, Navigation, and Control Subsystem (GN&CS) or manual commands from the astronauts.

The 16 thrust chamber assemblies (thrusters) and the propellant and helium sections that comprise the RCS are located in or on the ascent stage. (See figure 3-5.1.) The hypergolic propellants used in the RCS are identical with those used in the MPS. The fuel - Aerozine 50 - is a mixture of approximately 50% each of hydrazine and unsymmetrical dimethylhydrazine. The oxidizer is nitrogen tetroxide. In the LM, the injection ratio of oxidizer to fuel is approximately 2 to 1. The total RCS propellant weight is approximately 633 pounds.

The thrusters are small rocket engines, each capable of delivering 100 pounds of thrust. They are arranged in clusters of four, mounted on four outriggers equally spaced around the ascent stage. Each cluster is enclosed in a thermal shield that aids in maintaining a temperature-controlled environment for the propellant lines, minimizes heat loss, and reflects radiated engine heat and solar heat. In each cluster, two thrusters are mounted parallel to the vehicle X-axis, facing in opposite directions; the other two are spaced 90° apart, in a plane normal to the X-axis and parallel to the Y-axis and Z-axis. Four plume deflectors attached to the descent stage extend upward to the nozzle of each downward-firing TCA. The deflectors shield the descent stage structure from the downward exhaust plume.

The RCS is made up of two parallel, independent systems (A and B), which, under normal conditions, function together. Each system consists of eight thrusters, a helium pressurization section, and a propellant feed section. The two systems are interconnected by a normally closed crossfeed arrangement that enables the astronauts to operate all 16 thrusters from a single propellant supply. Complete attitude and translation control is therefore available even if one system's propellant supply is depleted or fails. Functioning alone, either RCS system can control the vehicle, although with slightly reduced efficiency.

In addition to the RCS propellant supply, the thrusters can use propellants from the ascent propulsion section. This method of feeding the thrusters, which requires the astronauts to open interconnect lines between the ascent tanks and RCS manifolds, is normally used only during periods of ascent engine thrusting. Use of ascent propulsion section propellants is intended to conserve RCS propellants, which may be needed during docking maneuvers.

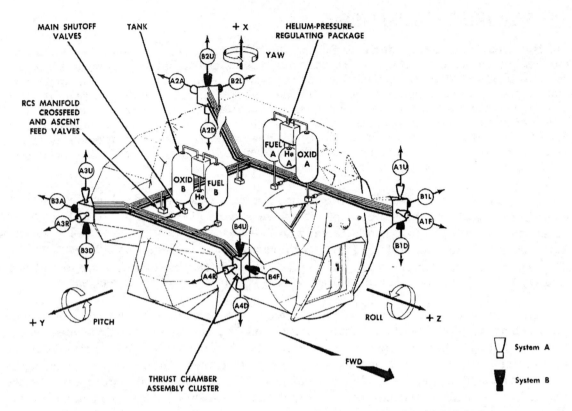

Figure 3-5.1. Reaction Control Subsystem – Major Equipment Location

3-5.1 THRUSTER SELECTION, OPERATION, AND CONTROL.

The GN&CS provides commands that select thrusters and fire them for durations ranging from a short pulse to steady-state operation. The thrusters can be operated in an automatic mode, attitude hold mode, or a manual override mode.

Normally, the RCS operates in the automatic mode; all navigation, guidance, stabilization, and steering functions are initiated and commanded by the LM guidance computer (primary guidance and navigation section) or the abort electronics assembly (abort guidance section).

The attitude hold mode is a semiautomatic mode in which either astronaut can institute attitude and translation changes. When an astronaut displaces his attitude controller, an impulse proportional to the amount of displacement is routed to the computer, where it is used to perform steering calculations and to generate the appropriate thruster-on command. When the astronaut returns his attitude controller to the neutral (detent) position, the computer issues a command to maintain attitude. For a translation maneuver, either astronaut displaces his thrust/translation controller. This sends a discrete to the computer to issue a thruster-on command to selected thrusters. When this controller is returned to neutral, the thrusters cease to fire.

If the abort guidance section is in control, attitude errors are summed with the proportional rate commands from the attitude controller and a rate-damping signal from the rate gyro assembly. The abort guidance equipment uses this data to perform steering calculations, which result in specific thruster-on commands. The astronauts can select two or four X-axis thrusters for translation maneuvers, and they can inhibit the four upward-firing thrusters during the ascent thrust phase, thus conserving propellants. In the manual mode, the four-jet hardover maneuver, instituted when either astronaut displaces his attitude controller fully against the hard stop, fires four thrusters simultaneously, overriding any automatic commands.

For an automatic MPS ullage maneuver, the astronauts select whether two or four downward-firing thrusters should be used. Firing two thrusters conserves RCS propellants; however, it takes longer to settle the MPS propellants. Under manual control, a pushbutton fires the four downward-firing thrusters continuously until the pushbutton is released.

3-5.2 PRESSURIZATION SECTION. (See figure 3-5.2.)

Because the two RCS systems are identical, only one system is described. The RCS propellants are pressurized with high-pressure gaseous helium, stored at ambient temperature. The helium tank outlet remains sealed by parallel-connected, redundant explosive valves until the astronauts prepare the RCS for operation. When the explosive valves are fired, helium enters the pressurization line and flows through a filter and a restrictor orifice that dampens the initial helium surge. The helium then flows through a pair of pressure regulators connected in series. The primary (upstream) regulator is set to reduce pressure to approximately 181 psia. The secondary (downstream) regulator is set for a slightly higher output. In normal operation, the primary regulator is in control and provides proper propellant tank pressurization.

The regulated helium flows into separate paths leading to the oxidizer and fuel tanks. Each flow path has quadruple check valves that permit flow in one direction only, thus preventing backflow of propellant vapors if seepage occurs in the propellant tank bladders. A relief valve assembly protects each propellant tank against overpressurization.

3-5.3 PROPELLANT FEED SECTION.

Fuel and oxidizer are contained in flexible bladders in the propellant tanks. The larger tanks contain oxidizer; the smaller tanks, fuel. Helium routed into the void between the bladder and the tank wall squeezes the bladder to positively expel the propellant under zero-gravity conditions. The propellants flow through normally open main shutoff valves into separate fuel and oxidizer manifolds that lead to the thrusters. A switch on the control panel enables the astronauts to simultaneously close a pair of fuel and oxidizer main shutoff valves, thereby isolating a system's propellant tanks from its thrusters. After shutting off one system, the astronauts can restore operation of all 16 thrusters by opening the crossfeed valves between the system A and B manifolds.

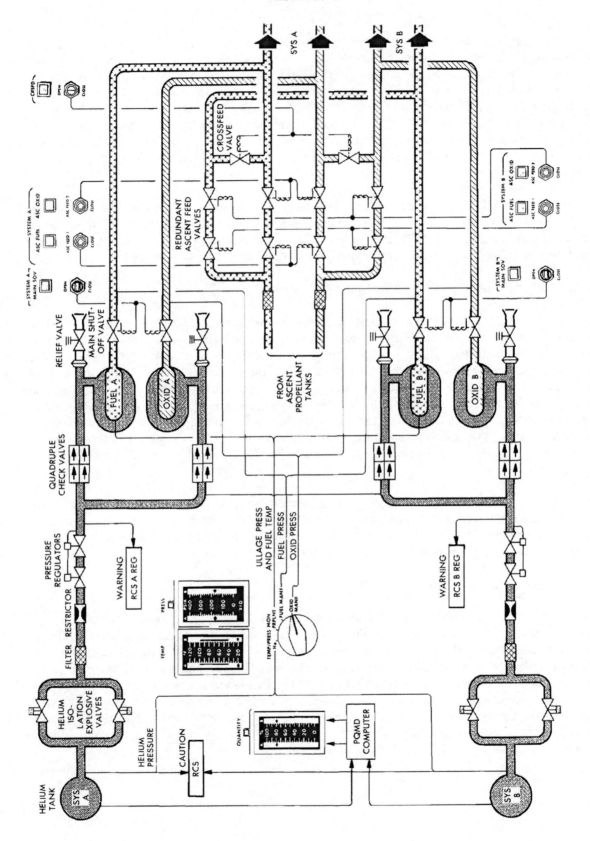

Figure 3-5.2. Helium Pressurization and Propellant Feed Sections – Flow Diagram

Transducers in the propellant tanks enable the astronauts to monitor helium pressure and fuel temperature. A propellant quantity measuring device (PQMD), consisting of a helium pressure/temperature probe and an analog computer, measures the total quantity of propellants in the fuel and oxidizer tanks. The output voltage of the analog computer is fed to an indicator and is displayed as percentage of propellant remaining in the tanks.

During ascent engine firing, the astronauts may open the normally closed ascent propulsion section/RCS interconnect lines. Control panel switches open the interconnect valves in fuel-oxidizer pairs for an individual RCS system, or for both systems simultaneously.

Fuel and oxidizer manifolds in each RCS system feed propellants to two thrusters in each cluster. (See figure 3-5.3.) In the LM, each of these thruster pairs could be shut off by switches that controlled a normally open pair of fuel and oxidizer isolation valves, permitting isolation of only two thrusters rather than disabling an entire system. When a thruster pair switch is in the closed position, it issues a signal informing the LM guidance computer that the related isolation valves are closed and that alternate thrusters must be selected. For the modified LM, the isolation valves have been removed. The LM guidance computer inhibit function, the telemetry function, and the caution and warning electronics assembly reset functions have been maintained by use of the LGC thruster pair command switches.

3-5.4 THRUSTER SECTION. (See figure 3-5.4.)

The RCS thrusters are radiation-cooled, pressure-fed, bipropellant rocket engines that operate in a pulse mode to generate short thrust impulses for fine attitude corrections or in a steady-state mode to produce continuous thrust for major attitude or translation changes. In the pulse mode, the thrusters are fired intermittently in bursts of less than 1-second duration - the minimum pulse may be as short as 14 milliseconds - however, the thrust level does not build up to the full 100 pounds that each thruster can produce. In the steady-state mode, the thrusters are fired continuously (longer than 1 second) to produce a stabilized 100 pounds of thrust until the shutoff command is received.

Two electric heaters, which encircle the thruster injector, control propellant temperature by heating the combustion chamber and the propellant solenoid valves. The heaters ensure that the combustion chambers are properly preheated for instantaneous thruster starts. The heaters normally operate in an automatic mode; redundant thermal switches (two connected in parallel for each thruster) sense injector temperature and turn the heaters on and off to maintain the temperature close to +140° F. The astronauts can determine, by use of a temperature indicator, if a cluster requires temperature correction. They would then override the automatic mode to restore the cluster temperature.

Propellants are prevented from entering the combustion chamber by the dual-coil, solenoid-operated shutoff valves at the thruster inlet ports. These valves are normally closed. When an automatic or a manual command energizes one of the coils, the valves open, permitting the propellants to flow through the injector into the combustion chamber where ignition occurs. The fuel valve opens 2 milliseconds before the oxidizer valve, to provide proper ignition characteristics. Orifices at the valve inlets meter the propellant flow so that the desired oxidizer-to-fuel mixture ratio is obtained.

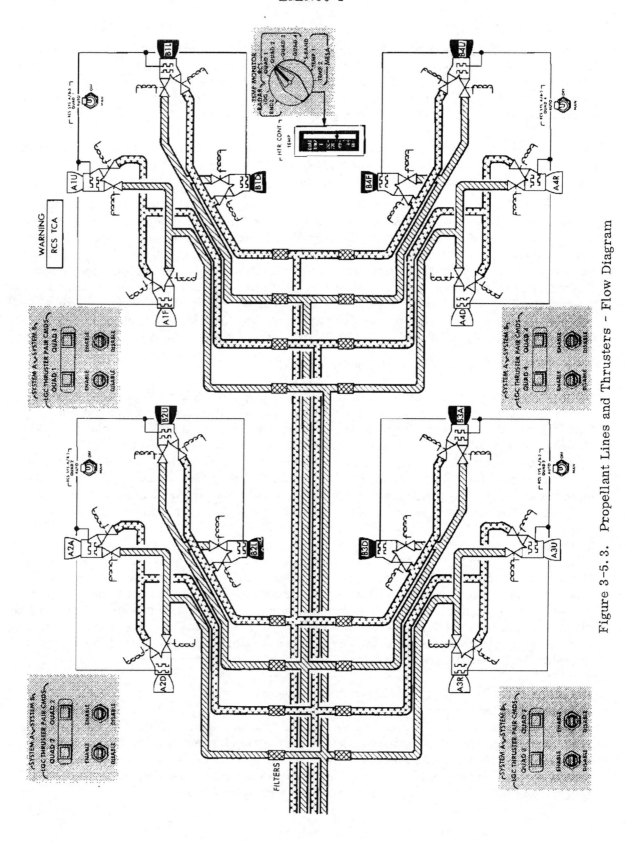

Figure 3-5.3. Propellant Lines and Thrusters – Flow Diagram

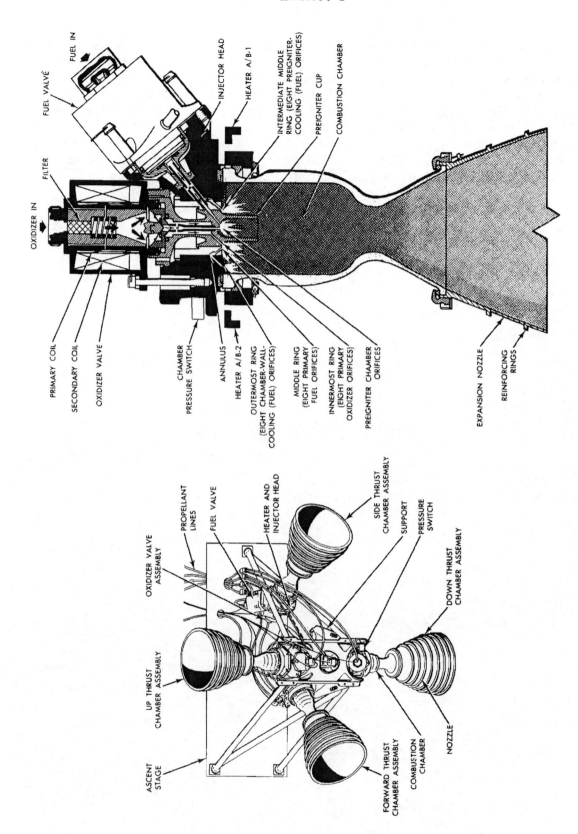

Figure 3-5.4. Thrust Chamber Assembly and Cluster

As the propellants mix and burn, the hot combustion gases increase the chamber pressure, accelerating the gas particles through the chamber exit. The gases are expanded through the divergent section of the nozzle at supersonic velocity, reaching a reactive force of 100 pounds of thrust in the vacuum of space. The gas temperature within the combustion chamber stabilizes at approximately 5,200° F. The temperature at the nonablative chamber wall is maintained at a nominal 2,800° F by a combined method of film cooling (a fuel stream sprayed against the wall) and radiation cooling (dissipation of heat from the wall surface into space). The combustion chamber is made of machined molybdenum, coated with silicon to prevent oxidation of the base metal.

When the thruster-off command is received, the coils in the propellant valves deenergize, and spring pressure closes the valves. Porpellant trapped in the injector is ejected and burned for a short time, while thrust decays to zero pounds.

3-6. <u>ELECTRICAL POWER SUBSYSTEM.</u>

The Electrical Power Subsystem (EPS) is the principal source of electrical power for the LM. (See figure 3-6.1.) Electrical power is supplied by four silver-zinc batteries in the descent stage and two in the ascent stage. For the modified vehicle, electrical power is supplied by five silver-zinc batteries in the descent stage and two in the ascent stage. The batteries provide dc for the EPS d-c section; two solid-state inverters supply the a-c section. Both sections supply operating power to respective electrical buses, which supply the subsystems through circuit breakers. Other batteries supply power to trigger explosive devices, to operate the portable life support system, and to operate scientific equipment.

Two descent stage batteries power the vehicle from T-30 minutes until after transposition and docking, at which time the vehicle receives electrical power from the CSM. After separation from the CSM, during the powered descent phase of the mission, four descent stage batteries are paralleled with the ascent stage batteries. Paralleling the batteries ensures the minimum required voltage for all possible vehicle operations. During lunar stay, specific combinations of 4 of the 5 descent stage batteries can be paralleled to provide vehicle power. Before lift-off from the lunar surface, ascent stage battery power is introduced, descent power is terminated, and descent feeder lines are deadfaced and severed. Ascent stage battery power is then used until after final docking and astronaut transfer to the CM. The batteries are controlled and protected by electrical control assemblies, a relay junction box, and a deadface relay box, in conjunction with the control and display panel.

In addition to being the primary source of electrical power for the vehicle during the mission, the EPS is the distribution point for externally generated power during prelaunch and docked operations. Prelaunch power is initially supplied from external ground power supplies until approximately T-7 hours. At this time, the vehicle ground power supply unit is removed and d-c power from the launch umbilical tower is connected. From T-30 minutes until transposition and docking, the EPS supplies internal d-c power. After docking, vehicle power is shut down and the CSM supplies d-c power to the vehicle. Before vehicle separation, all vehicle internally supplied electrical power is restored.

3-6.1. FUNCTIONAL DESCRIPTION. (See figure 3-6.3.)

The outputs of the five descent stage batteries and two ascent stage batteries are applied to four electrical control assemblies. The two descent stage electrical control assemblies provide control circuits for each descent battery. A battery relay box is provided to accomodate the fifth (lunar stay) battery to the four descent battery control circuits. The two ascent stage electrical control assemblies provide four battery control circuits, two control circuits for each ascent battery. The electrical control assembly monitors reverse-current, overcurrent, and overtemperature within each battery. Each control circuit can detect a bus or feeder short. If an overcurrent condition occurs in a descent or ascent battery, the control circuit operates a main feed contactor associated with the malfunctioning battery to remove the battery from the distribution system.

Ascent and descent main power feeders are routed through circuit breakers to the d-c buses. From these buses, power is distributed through circuit breakers to all subsystems. The two inverters, which make up the a-c section power source, are connected to either of two a-c buses. Either inverter, when selected, can supply the vehicle a-c requirements.

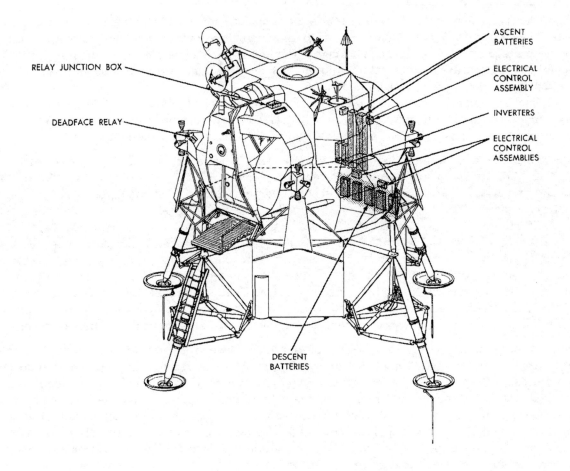

Figure 3-6.1. Electrical Power Subsystem - Major Equipment Location

Throughout the mission, the astronauts monitor the primary a-c and d-c voltage levels, d-c current levels, and the status of all main power feeders. The electrical power control and indicator panel in the cabin has talkbacks that indicate main power feeder status, indicators that display battery and bus voltages and currents, and component caution lights. The component caution lights are used to detect low bus voltages; out-of-limit, a-c bus frequencies; and battery malfunctions. Backup a-c and d-c power permits the astronauts to disconnect, substitute, or reconnect batteries, feeder lines, buses, or inverters to assure a continuous electrical supply.

3-6.2. EQUIPMENT.

3-6.2.1. Descent Stage Batteries. The five descent stage batteries are composed of silver-zinc plates, with a potassium hydroxide electrolyte. Each battery has 20 cells, weighs 135 pounds, and has a 400-ampere-hour capacity (25 amperes at 28 volts dc for 16 hours, at +80° F). The batteries operate in a vacuum while cooled by an Environmental Control Sybsystem (ECS) cold rail assembly to which the battery heat sink surface is mounted. Five thermal sensors monitor cell temperature limits (+145°±5° F) within each battery; an overtemperature condition causes a caution light to go on. The batteries initially have high-voltage characteristics; a low-voltage tap is provided at the 17th cell of two batteries for use from T-30 minutes through transposition and docking. The high-voltage taps are used for all other normal vehicle operations. If one descent stage battery fails during the mission, the remaining four descent stage batteries can provide sufficient power to complete a curtailed lunar surface mission.

3-6.2.2. Ascent Stage Batteries. The ascent stage batteries are composed of silver-zinc plates, with a potassium hydroxide electrolyte. Each battery weighs 125 pounds, and has a 296-ampere-hour capacity (50 amperes at 28 volts for 5.9 hours, at +80° F). To provide independent battery systems, the batteries are normally not paralleled during the ascent phase of the mission. The batteries operate in a vacuum while cooled by ECS cold rails to which the battery heat sink surface is mounted. The nominal operating temperature of the batteries is approximately +80° F. Battery temperature in excess of +145°±5° F closes a thermal sensor, causing a caution light to go on. The batteries supply d-c power from normal staging to final docking of the ascent stage with the orbiting CSM or in the event of a malfunction that requires separation of the ascent and descent stages. If one ascent stage battery fails, the remaining battery provides sufficient power to accomplish safe rendezvous and docking with the CSM during any part of the mission.

3-6.2.3. Descent Stage Electrical Control Assemblies. The two descent stage electrical control assemblies control and protect the descent stage batteries. Each assembly has a set of control circuits for each power source accommodated. A failure in one set of control circuits does not affect the other set. The protective circuits of the assembly automatically disconnect a descent stage power source if an overcurrent condition occurs and cause a caution light to go on if a battery overcurrent, reverse-current, or overtemperature condition is detected.

The major elements of each assembly are high- and low-voltage main feed contactors, current monitors, overcurrent relays, reverse-current relays, and power supplies. An auxiliary relay supplies system logic contact closures to other control assemblies in the power distribution system.

The reverse-current relay causes a caution light to go on when current flow in the direction opposite to normal current flow exceeds 10 amperes for at least 5 seconds. Unlike the overcurrent relay, the reverse-current relay does not open the related main feed contactor and is self-resetting when the current monitor ceases to detect a reverse-current condition. During reverse-current conditions, the related contactor must be manually switched open. The control assembly power supplies provide ac for current-monitor excitation and regulated dc for the other circuits.

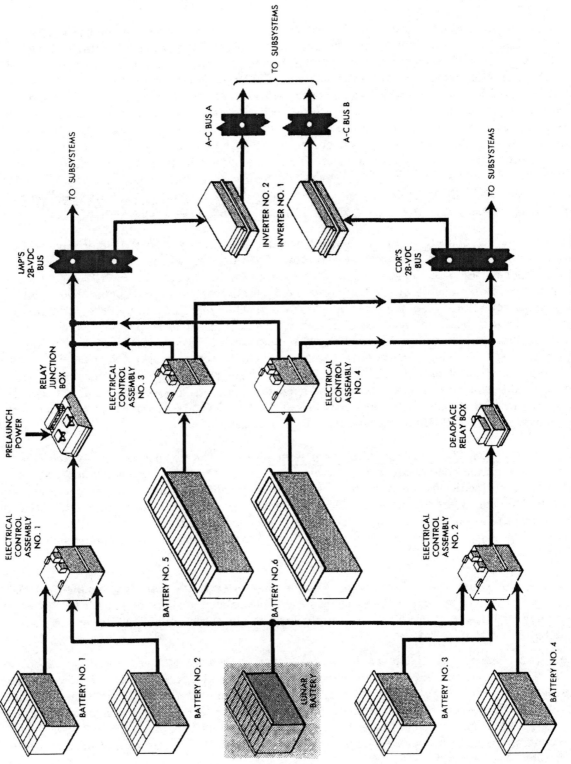

Figure 3-6.2. Electrical Power Subsystem – Flow Diagram

3-6.2.4. <u>Ascent Stage Electrical Control Assemblies.</u> The two ascent stage electronic control assemblies individually control and protect the two ascent stage batteries in nearly the same manner as the descent stage control assemblies. Each assembly contains electrical power feed contactors, an overcurrent relay, a reverse-current relay, and a current monitor. Each ascent stage battery can be connected to its normal or backup main feeder line via the normal or the backup main feed contactor in its respective assembly. Both batteries are thereby connected to the primary d-c power buses. The normal feeder line has overcurrent protection; the backup feeder line does not.

3-6.2.5. <u>Relay Junction Box.</u> The relay junction box provides the following:

Control logic and junction points for connecting external prelaunch power (via the launch umbilical tower) to the LM Pilot's d-c bus.

Control and power junction points for connecting descent stage and ascent stage electrical control assemblies to the LM Pilot's d-c bus.

Deadfacing (electrical isolation) of half of the power feeders between the descent and ascent stages.

The relay junction box controls all low-voltage contactors (on and off) from the launch umbilical tower and CSM, and all low- and high-voltage descent power contactors (off) on receipt of an abort stage command. The junction box includes abort logic relays, which, when energized by an abort stage command, close the ascent stage battery main feed contactors and open the deadface main feed contactors and deadface relays. The deadface relay is manually opened and closed or automatically opened when the abort logic relays close. The deadface relay in the junction box deadfaces half of the main power feeders between the descent and ascent stages; the other half of the power feeders is deadfaced by the deadface relay in the deadface relay box. The ascent stage batteries then provides primary d-c power to the vehicle.

3.6.2.6. <u>Deadface Relay Box.</u> The deadface relay box deadfaces those power feeders that are not controlled by the relay junction box, in the same manner as the relay junction box. Two individual deadfacing facilities (28 volts for each circuit breaker panel) are provided.

3-6.2.7. <u>Inverters.</u> Two identical redundant, 400-Hz inverters individually supply the primary a-c power required. Inverter output is derived from a 28-volt d-c input. The output of the inverter stage is controlled by 400-Hz pulse drives developed from a 6.4-kilopulse-per-second (kpps) oscillator, which is, in turn, synchronized by timing pulses from the Instrumentation Subsystem. An electronic tap changer sequentially selects the output of the tapped transformer in the inverter stage, converting the 400-Hz square wave to an approximate sine wave of the same frequency. A voltage regulator maintains the inverter output at 115 volts ac during normal load conditions by controlling the amplitude of a dc-to-dc converter output. The voltage regulator also compensates for variations in the d-c input and a-c output load. When the voltage at a bus is less than 112 volts ac, or the frequency is less than 398 Hz or more than 402 Hz, a caution light goes on. The light goes off when the malfunction is remedied.

3-6.2.8. **Battery Relay Box.** The battery relay box adapts the lunar stay battery to the descent electrical control assemblies. It inhibits battery overtemperature monitoring when the lunar stay battery is not in use, switches current monitor signals and malfunction indications to the appropriate monitors, and it inhibits the simultaneous application of lunar stay battery voltage to both lunar buses. It also provides a fused circuit for lunar stay battery voltage measurements.

3-6.2.9. <u>Circuit Breaker and EPS Control Panels.</u> All primary a-c and d-c power feed circuits are protected by circuit breakers on the Commander's and LM Pilot's buses. The two d-c buses are electrically connected by the main power feeder network. Functionally redundant equipment is placed on both d-c buses (one on each bus), so that each bus can individually perform a mission abort.

3-6.2.10. <u>Sensor Power Fuse Assemblies.</u> Two sensor power fuse assemblies, in the aft equipment bay, provide a secondary d-c bus system that supplies excitation to transducers in other subsystems that develop display and telemetry data. During prelaunch procedures, primary power is supplied to the assemblies from the Commander's 28-volt d-c bus. Before launch, power from the launch umbilical tower is disconnected, and power is supplied to the sensor power fuse assemblies from the LM Pilot's 28-volt d-c bus. Each assembly comprises a positive d-c bus, negative return bus, and 40 fuses. All sensor return lines are routed to a common ground bus.

3-7. ENVIRONMENTAL CONTROL SUBSYSTEM. (See figure 3-7.1.)

The Environmental Control Subsystem (ECS) enables pressurization of the cabin and space suits, controls the temperature of electronic equipment, and provides breathable oxygen for the astronauts. It also provides water for drinking, cooling, fire extinguishing, and food preparation, and supplies oxygen and water to the portable life support system (PLSS).

The major portion of the ECS is in the cabin. The peripheral ECS equipment, such as oxygen and water tanks, is located outside the cabin, in the ascent and descent stages. The ECS consists of the following sections:

Atmosphere revitalization section (ARS)

Oxygen supply and cabin pressure control section (OSCPCS)

Water management section (WMS)

Heat transport section (HTS)

The ARS purifies and conditions the oxygen for the cabin and the space suits. Oxygen conditioning consists of removing carbon dioxide, odors, particulate matter, and excess water vapor.

The OSCPCS stores gaseous oxygen and maintains cabin and suit pressure by supplying oxygen to the ARS to compensate for crew metabolic consumption and cabin or suit leakage. Two oxygen tanks in the descent stage provide oxygen during descent and lunar stay. Two oxygen tanks in the ascent stage are used during ascent and rendezvous.

The WMS supplies water for drinking, cooling, fire extinguishing, and food preparation, and for refilling the PLSS cooling water tank. It also provides for delivery of water from ARS water separators to HTS sublimators and from water tanks to ARS and HTS sublimators.

The water tanks are pressurized with nitrogen before launch, to maintain the required pumping pressure in the tanks. Two descent stage tanks supply most of the water required until staging occurs. After staging, water is supplied by the two ascent stage tanks. A self-sealing valve delivers water for drinking and food preparation.

The HTS consists of a primary coolant loop and a secondary coolant loop. The secondary loop serves as a backup loop; it functions if the primary loop fails. A water-glycol solution (coolant) circulates through each loop. The primary loop provides temperature control for batteries, electronic equipment that requires active thermal control, and for the oxygen that circulates through the cabin and space suits. The batteries and electronic equipment are mounted on cold plates and rails through which coolant is routed to remove waste heat. The cold plates used for equipment that is required for mission abort contain two separate coolant passages: one for the primary loop and one for the secondary loop. The secondary coolant loop, which is used only if the primary loop is inoperative, serves only these cold plates.

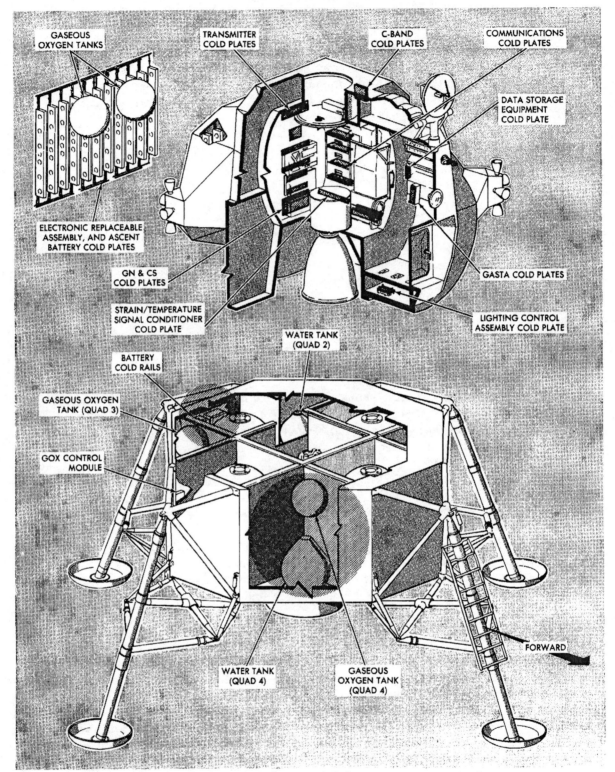

GASEOUS
OXYGEN TANKS

TRANSMITTER
COLD PLATES

C-BAND
COLD PLATES

COMMUNICATIONS
COLD PLATES

DATA STORAGE
EQUIPMENT
COLD PLATE

ELECTRONIC REPLACEABLE
ASSEMBLY, AND ASCENT
BATTERY COLD PLATES

GN & CS
COLD PLATES

GASTA COLD PLATES

STRAIN/TEMPERATURE
SIGNAL CONDITIONER
COLD PLATE

LIGHTING CONTROL
ASSEMBLY COLD PLATE

WATER TANK
(QUAD 2)

BATTERY
COLD RAILS

GASEOUS OXYGEN
TANK (QUAD 3)

GOX CONTROL
MODULE

FORWARD

WATER TANK
(QUAD 4)

GASEOUS
OXYGEN TANK
(QUAD 4)

Figure 3-7.1. Environmental Control Subsystem - Component Location
(Sheet 1 of 2)

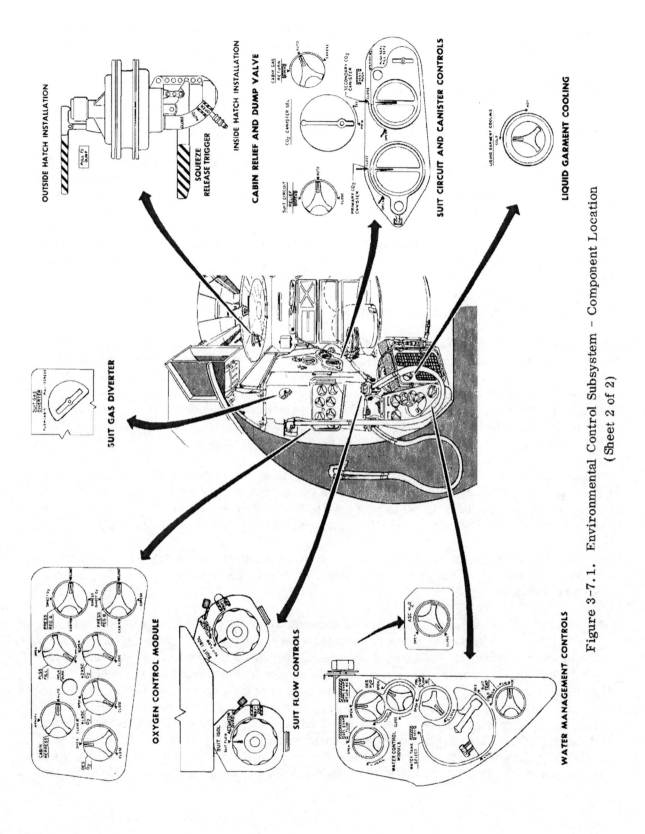

Figure 3-7.1. Environmental Control Subsystem – Component Location (Sheet 2 of 2)

In-flight waste heat rejection from both coolant loops is achieved by the primary and secondary sublimators, which are vented overboard. A coolant pump recirculation assembly contains all the HTS coolant pumps and associated check and relief valves. Coolant flow from the assembly is directed through parallel circuits to the cold plates for the electronic equipment and the oxygen-to-glycol heat exchangers in the ARS.

3-7.1. ATMOSPHERE REVITALIZATION SECTION.

The ARS is a recirculation system that conditions oxygen by cooling or heating, dehumidifying, and deodorizing it for use within the space suits and cabin, and circulates water through the liquid cooling garment to provide cooling during peak heat loads. The major portion of the ARS is within the suit circuit assembly.

In normal operation, conditioned oxygen flows to the space suits and is discharged through the return umbilical to the suit circuit. Suit circuit pressure, sensed at a point downstream of the suits, is referenced to the oxygen regulators that control pressure by supplying make-up oxygen to the suit circuit. The suit circuit relief valve protects the suit circuit against overpressurization, by venting the cabin.

The cabin position of the suit gas diverter valve is used during pressurized-cabin operation, to direct sufficient conditioned oxygen to the cabin to control carbon dioxide and humidity levels. Pulling the valve handle selects the egress position to isolate the suit circuit from the cabin. The egress position is used for all unpressurized-cabin operations, as well as the closed-suit mode with pressurized cabin. An electrical solenoid override automatically repositions the valve from cabin to egress when cabin pressure drops below the normal level or when the egress position of the pressure regulators is selected.

With the suit gas diverter valve set to the cabin position, cabin discharge oxygen is returned to the suit circuit through the cabin gas return valve. Setting the cabin gas return valve to the automatic position enables cabin pressure to open the valve. When the cabin is depressurized, differential pressure closes the valve, preventing suit pressure loss.

A small amount of oxygen is tapped from the suit circuit upstream of the PGA inlets and fed to the carbon dioxide partial pressure sensor, which provides a voltage to the CO_2 indicator, pressure indicator.

The primary and secondary carbon dioxide and odor removal canisters are connected to form a parallel loop. The primary canister contains a LM cartridge with a capacity of 41 man-hours; the secondary canister, a PLSS cartridge with a capacity of 14 man-hours. A debris trap in the primary canister cover prevents particulate matter from entering the cartridge. A relief valve in the primary canister permits flow to bypass the debris trap if it becomes clogged. Oxygen is routed to the carbon dioxide and odor removal canisters through the canister selector valve. The carbon dioxide level is controlled by passing the flow across a bed of lithium hydroxide (LiOH); odors are removed by adsorption on activated charcoal. When carbon dioxide partial pressure reaches or exceeds 7.6 mm Hg, as indicated on the partial pressure CO_2 indicator, the CO_2 component caution light and ECS caution light go on. The CO_2 canister selector valve is then set to the secondary position, placing the secondary canister onstream. The primary cartridge is replaced and the CO_2 canister selector valve is set to the primary position, placing the primary canister back onstream. (The CO_2 component caution light also goes on when the CO_2 canister selector valve is in the secondary position.)

From the canisters, conditioned oxygen flows to the suit fan assembly, which maintains circulation in the suit circuit. Only one fan operates at a time. The ECS suit fan 1 circuit breaker is closed and the SUIT FAN selector switch is set to 1 to initiate suit fan operation. At startup, a fan differential pressure sensor is in the low position (low ΔP), which, through the fan condition signal control, energizes the ECS caution light and suit fan component caution light. The lights remain on until the differential pressure across the operating fan increases sufficiently to cause the differential pressure sensor to move to the normal position. If the differential pressure drops to 6.0 inches of water or less, the lights go on and switch-over to fan No. 2 is required. The ECS caution light goes off when fan No. 2 is selected and the suit fan warning light goes on. The suit fan component caution light goes off when fan No. 2 comes up to speed and builds up normal differential pressure. The suit fan warning light and suit fan component caution light go off if fan No. 2 differential pressure reaches 9.0 inches of water. The fan check valve permits air to pass from the operating fan without backflow through the inoperative fan.

From the check valve, the conditioned oxygen passes through a sublimator to the cooling heat exchanger. The sublimator cools the oxygen under emergency conditions. Heat transfer to the coolant in the heat exchanger reduces gas temperature and causes some condensation of water vapor.

Conditioned oxygen is next routed to two parallel-redundant water separators through the water separator selector valve. One separator, selected by pushing or pulling the water separator selector valve handle, is operated at a time. The separator is driven by the gas flowing through it. Moisture removed from the oxygen is discharged under a dynamic head of pressure sufficient to ensure positive flow from the separator to the WMS. A water drain carries some water from the separators to a surge (collection) tank outside the recirculation system.

Conditioned oxygen from the water separator is mixed with makeup oxygen from the OSCPCS to maintain system pressure. The mixture flows through the regenerative (heating) heat exchanger, where the temperature may be increased to the suit isolation valves. The suit temperature control valve on the water control module controls the flow of coolant through the regenerative heat exchanger. Setting the valve to the increase hot position increases oxygen temperature; setting it to decrease cold position reduces the temperature.

3-7.2. SUIT LIQUID COOLING ASSEMBLY.

The suit liquid cooling assembly assists in removing metabolic heat by circulating cool water through the liquid cooled garment (LCG). A pump maintains the flow of warm water returning from the LCG through the water umbilicals. An accumulator in the system compensates for volumetric changes and leakage. A mixing bypass valve controls the quantity of water that flows through the water-glycol heater exchanger. This bypassed (warm) water is mixed with the cool water downstream of the heat exchanger and flows through the water umbilicals back to the LCG.

3-7.3.   OXYGEN SUPPLY AND CABIN PRESSURE CONTROL SECTION. (See figure 3-7.2.)

The ECS descent stage oxygen supply hardware consists of two descent oxygen tanks (in quads 3 and 4), a high-pressure fill coupling, a high-pressure oxygen control assembly, an interstage flex line, and a descent stage disconnect. The modified LM has redundant PLSS oxygen fill lines. One fill line, directly from a single descent tank, has its own regulator and overboard relief valves. The other fill line enables filling of the PLSS from the ascent stage tanks. The descent tank pressure transducer, part of the Instrumentation Subsystem (IS), generates an output proportionate to tank pressure.

The ascent stage oxygen supply hardware consists of an ascent stage disconnect, interstage flex line, oxygen module, two ascent oxygen tanks, and the cabin pressure switch. Two automatic cabin pressure relief and dump valves, one in each hatch, are provided to protect the cabin from overpressurization. Two ascent stage tank pressure transducers and a selected oxygen supply transducer, part of the IS, operate in conjunction with the OSCPCS.

The OSCPCS stores gaseous oxygen, replenishes the ARS oxygen, and provides refills for the PLSS oxygen tank. Before staging, oxygen is supplied from the descent stage oxygen tanks.

After staging, or if the descent tanks are depleted, the ascent stage oxygen tanks supply oxygen to the oxygen control module. The high-pressure assembly in the descent stage, and the oxygen control module in the ascent stage, contain the valves and regulators necessary to control oxygen in the OSCPCS.

A high-pressure regulator reduces descent tank pressure, approximately 2,730 psia, to a level that is compatible with the components of the oxygen control module, approximately 900 psig. A series-redundant overboard relief valve protects the oxygen control module against excessive pressure caused by a defective regulator or by flow through the bypass relief valve. If the pressure on the outlet side of the regulator rises to a dangerous level, the burst diaphragm assembly vents the high-pressure assembly to ambient. A poppet in the burst diaphragm assembly reseats when pressure in the high-pressure assembly is reduced to approximately 1,000 psig. Descent oxygen flow through the interstage disconnect to the oxygen control module is controlled with the descent oxygen shutoff valve. The interstage disconnect acts as a redundant seal to prevent loss of oxygen overboard after staging.

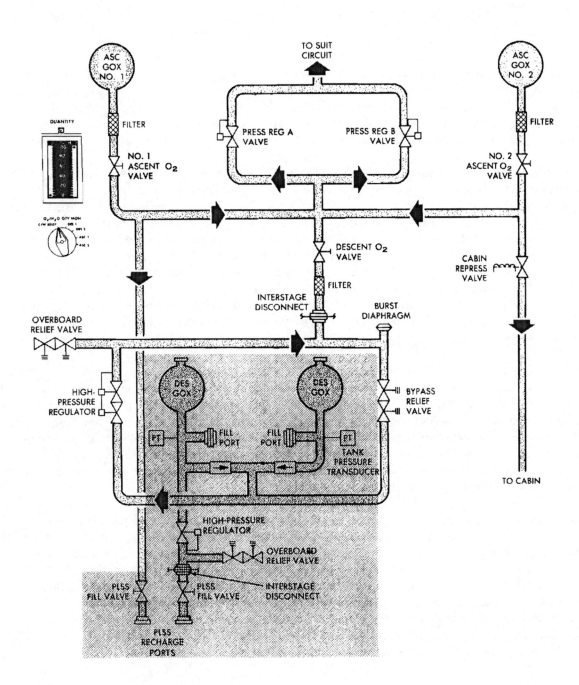

Figure 3-7.2. Oxygen Supply and Cabin Pressure
Control Section - Flow Diagram

When ascent stage oxygen is required, the ascent oxygen shutoff valves are used to select their respective tank. A mechanical interlock prevents the valves from being opened unless the descent oxygen shutoff valve is closed. The mechanical interlock may be overridden (if the descent oxygen shutoff valve cannot be closed and the ascent oxygen shutoff valves must be opened) by pressing the interlock override pushbutton on the oxygen control module.

From the oxygen shutoff valves, oxygen is routed to oxygen demand regulators, and a cabin repressurization and emergency oxygen valve. The oxygen demand regulators maintain the pressure of the suit circuit at a level consistent with normal requirements. Both regulators are manually controlled with a four-position handle; both are ordinarily set to the same position. The CABIN position is selected during normal pressurized cabin operations, to provide oxygen at 4.8 ± 0.2 psia. Setting the regulators to the egress position maintains suit circuit pressure at 3.8 ± 0.2 psia. The direct O_2 position provides an unregulated flow of oxygen into the suit circuit. The close position shuts off all flow through the regulator. In the cabin and egress positions, the regulator is internally modulated by a reference pressure from the suit circuit. The demand regulators are redundant; either one can fulfill the ARS oxygen requirements.

If both demand regulators are set to the cabin or direct O_2 position and cabin pressure drops to 3.7 to 4.45 psia, the cabin pressure switch energizes the cabin repressurization valve and oxygen flows through the valve into the cabin. If cabin pressure builds up to 4.45 to 5.0 psia, the cabin pressure switch deenergizes the valve solenoid, shutting off the oxygen flow. The valve can maintain cabin pressure at 3.5 psia for at least 2 minutes following a 0.5-inch-diameter puncture of the cabin. It responds to signals from the cabin pressure switch during pressurized-cabin operation and to a suit circuit pressure switch during unpressurized operation. Manual override capabilities are provided.

Both cabin relief and dump valves (one in the forward hatch, the other in the overhead hatch) are manually and pneumatically operated. They prevent excessive cabin pressure and permit deliberate cabin depressurization. The valves automatically relieve cabin pressure when the cabin-to-ambient differential reaches 5.4 to 4.8 psid. When set to the automatic position, the valves can be manually opened with their external handle. The valve in the overhead hatch can dump cabin pressure from 5.0 to 0.08 psia in 180 seconds without cabin inflow. The valve in the forward hatch requires 310 seconds to dump the same amount of cabin pressure because of the flow restriction caused by the bacteria filter. In addition to relieving positive pressure, the valves relieve a negative cabin pressure condition.

To egress from the LM, the oxygen demand regulators are set to the egress position, turning off the cabin fans and closing the suit gas diverter valve; the cabin gas return valve is set to the egress position; and cabin pressure is dumped by opening the cabin relief and dump valve. The bacteria filter on the cabin side of the cabin relief and dump valve in the forward hatch removes 95% of all bacteria larger than 0.5 micron from the dumped oxygen. When repressurizing the cabin, the cabin relief and dump valve is set to the automatic position, the oxygen demand regulator valves are set to the cabin position, and the cabin gas return valve is set to the automatic position. The cabin warning light goes on when the regulators are set to the cabin position and goes off when cabin pressure reaches the actuation pressure of the cabin pressure switch.

3-7.4. WATER MANAGEMENT SECTION. (See figure 3-7.3.)

The WMS stores water for metabolic consumption, evaporative cooling, fire extinguishing, and PLSS water tank refill. It controls the distribution of this stored water and the water reclaimed from the ARS by the water separators. Reclaimed water is used only for evaporative cooling, in the ECS sublimators. Water is stored in two tanks in the descent stage and two identical smaller tanks in the upper midsection of the ascent stage. All four tanks are bladder-type vessels, which are pressurized with nitrogen before launch. The controls for the WMS are grouped together on the water control module located to the right rear of the LM Pilot's station.

Water from the descent stage water tanks is fed through a manually operated shutoff valve and check valves to the PLSS water disconnect. Water quantity remaining is determined by water quantity measuring devices, which sense the temperature/pressure ratio of the tank-pressurizing nitrogen, computes the volume of the water, and generates an output analogous to that volume. The output is displayed on the H_2O quantity indicator after the O_2/H_2O quantity monitor selector switch is set to the descent position. If the quantity of water in either of the descent tanks drops to below the required level, the water quantity caution light goes on. When the descent H_2O valve is opened, high-pressure water is available for drinking, food preparation, PLSS fill, and fire extinguishing.

When the vehicle is staged, the descent interstage water feed line is severed by the interstage umbilical guillotine, and water is supplied from the ascent stage water tanks. Water quantity in either ascent water tank is monitored as required by switching. The water quantity caution light goes on when a less-than-full condition exists in either tank or when the tank water levels are unequal. Water from ascent stage water tank No. 1 is fed through the ascent water valve for drinking, food preparation, and fire extinguishing. The PLSS is recharged from the descent water tanks.

Water from the four water tanks enter the water tank selector valve, which consists of two water-diverting spools. Setting the valve to the descent or ascent position determines which tank is on-line.

When using the descent tanks, water is supplied to the primary manifold (which consists of the primary pressure regulators and the primary evaporator flow No. 2 valve) by setting the

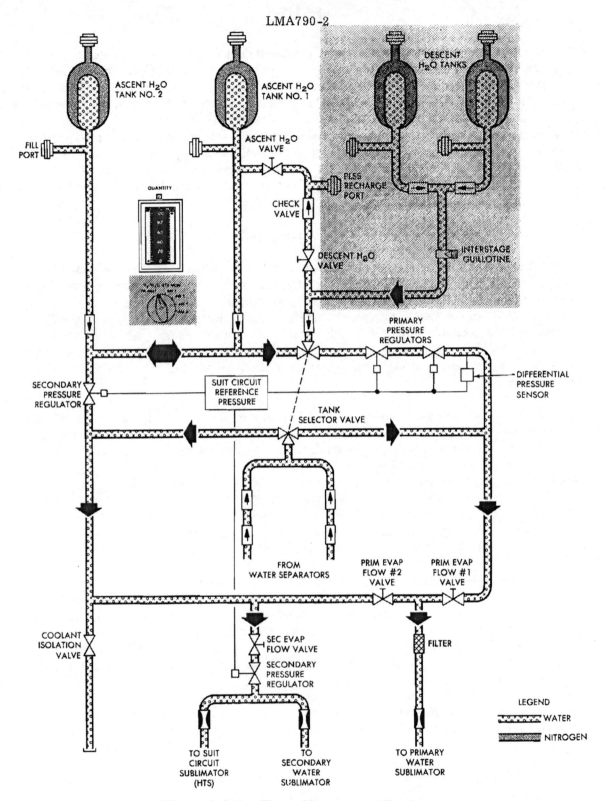

Figure 3-7.3. Water Management Section

water tank select valve to the descent position. The water flows through the series primary pressure regulators, which control water discharge pressure, in response to suit circuit gas reference pressure, at 0.5 to 1.0 psi above this gas pressure. With the primary evaporator flow valve opened, the water is routed to the primary sublimator. Discharge water from the water separator is routed through the secondary spool of the selector valve and joins the water from the primary pressure regulators. Setting the selector valve to ASC routes water from the ascent tanks through the primary pressure regulators and, with the primary evaporator flow No. 1 valve opened, to the primary sublimator. Water flow from the water separators is not changed by selection of the ASC position. If the primary pressure regulators fail, an alternative path to the primary sublimator is provided with the primary evaporator flow No. 2 valve opened. Water then flows directly from the ascent water tanks through the secondary pressure regulator and the primary evaporator flow No. 2 valve to the primary sublimator.

Under emergency conditions (failure of the primary HTS loop), water from the ascent tanks is directed through the secondary manifold (secondary pressure regulator) to the secondary sublimator and the suit circuit sublimator by opening the secondary evaporator flow valve. Discharge water from the water separators is also directed to the sublimator.

3-7.5. HEAT TRANSPORT SECTION. (See figure 3-7.4.)

The HTS consists of two closed loops (primary and secondary) through which a water-glycol solution is circulated to cool the suit circuits, cabin atmosphere, and electronic equipment. Coolant is continuously circulated through cold plates and cold rails to remove heat from electronic equipment and batteries. For the purpose of clarity, the primary and secondary coolant loops, and the primary and secondary coolant loop cold plates and rails are discussed separately in the following paragraphs. When necessary, the primary loop is also a heat source for the cabin atmosphere. Heat generated in the cabin is removed by the water-glycol and is rejected to space by sublimation.

The primary coolant loop is charged with coolant at the fill points and is then sealed. The glycol pumps force the coolant through the loop. The glycol accumulator maintains a constant head of pressure (5.25 to 9 psia, depending on coolant level) at the inlets of the primary loop glycol pumps. Coolant temperature at the inlets is approximately +40°F. A switch in a low-level sensor trips when only 10% ±5% of coolant volume remains in the accumulator. When tripped, the switch provides a telemetry signal and causes the glycol caution light to go on.

The coolant is routed to the pump through a filter. The pump is started by closing the appropriate circuit breakers and setting the glycol selector switch to pump 1 or pump 2. If the operating pump does not maintain a minimum differential pressure (ΔP) of 7±2 psi, the ΔP switch generates a signal to energize the ECS caution light and the glycol component caution light. Selecting the other primary pump deenergizes the lights when the onstream pump develops a minimum ΔP of 5.0 to 9.0 psi. If both primary pumps fail, the secondary loop is activated by setting the water tank selector valve to the secondary position, setting the glycol pump switch to INST (SEC), and closing the glycol pump secondary circuit breaker. Automatic transfer to primary pump No. 1 from primary pump No. 2 is initiated by closing the glycol

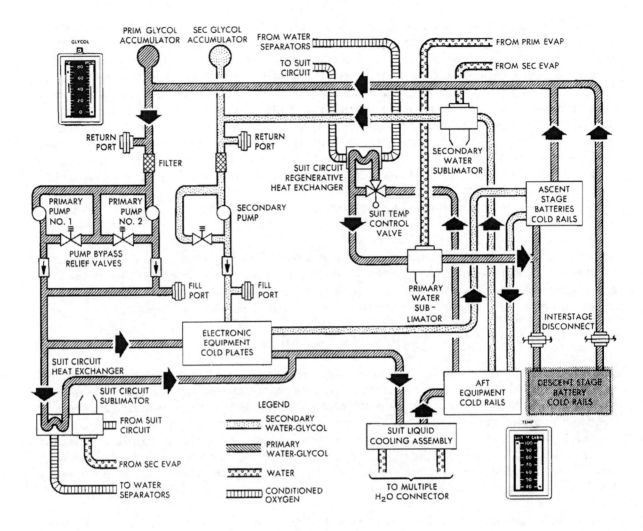

Figure 3-7.4. Heat Transport Section - Flow Diagram

automatic transfer circuit breaker and setting the selector switch to pump 1. When transfer is necessary, the caution lights go on, the transfer is accomplished, and the ECS caution light goes off.   The glycol pump component caution light remains on.

If primary loop ΔP exceeds 33 psi, the pump bypass relief valve opens and routes the coolant back to the pump inlet, relieving the pressure.   The valves start to open at 33 psi, are fully open at a maximum of 39 psia, and reseat at a minimum of 32 psia.   Check valves prevent coolant from feeding back through an inoperative primary pump.

Part of the coolant leaving the recirculation assembly flows to the suit circuit heat exchanger to cool the suit circuit gas of the ARA.   The remainder of the coolant flows to the electronic equipment mounted on cold plates.   The flow paths then converge and the coolant is directed to the liquid cooling garment water-glycol heat exchanger to cool suit water as required. The coolant then flows through the aft equipment bay cold rails.

A portion of the warmer coolant flow can be diverted to the suit circuit heat exchanger through the suit temperature control valve to increase suit inlet gas temperature. The diverted flow returning from the heat exchanger, combined with the bypassed coolant, is routed to the primary sublimator.

The sublimator reduces the temperature of the coolant by rejecting heat to space through sublimation of water, followed by venting of generated steam through an overboard duct. Deflector plates, attached to the duct, prevent escaping steam from applying thrust to the vehicle. Water is fed to the sublimator at a pressure that exceeds 4.0 psia, but is less than 6.5 psia. The water pressure must be less than the suit circuit static pressure plus the head pressure from the water separators to the sublimator. The water regulators, referenced to suit circuit pressure, are in the water feed line to the sublimator. Regulated water pressure varies from 0.5 to 1.0 psid above suit circuit pressure. The sublimator inlet-outlet temperatures are sensed by temperature transducers, which provide telemetry signals. Coolant from the sublimator flows through the ascent and descent battery cold rails, then returns to the recirculation assembly.

Two self-sealing disconnects upstream and downstream of the glycol pumps permit servicing of the HTS. Interstage disconnects are in coolant lines that connect to the descent stage. Before staging, coolant flows through the ascent and descent stage battery cold rails. After staging, the interstage disconnects separate, the lines are sealed by spring-loaded check valves, and the full coolant flow enters the ascent stage battery cold rails.

The secondary (emergency) coolant loop provides thermal control for those electronic assemblies and batteries whose performance is necessary to effect a safe return to the CSM. Cooling is provided by the secondary sublimator.

As in the primary loop, a secondary glycol accumulator provides pressure for the pump inlet side and compensates for loss due to leakage. A pump bypass relief valve relieves excessive pressure by routing coolant back to the pump inlet. A check valve at the discharge side of the glycol pump prevents coolant flow from bypassing the HTS during GSE operation. The coolant from the pump passes through the check valve to the secondary passage of the cold plates and cold rails of the electronics and batteries cold plate section. Waste heat is absorbed by the coolant. The warm coolant then flows to the secondary sublimator.

The secondary sublimator operates in the same manner as the primary sublimator in the primary coolant loop. Water for the sublimator is provided when the secondary evaporator flow valve is opened. The coolant returns to the pump for recirculation.

Equipment essential for mission abort is mounted on cold plates and rails that have two independent coolant passages, one for the primary loop and one for the secondary loop.

3-7.5.1. Primary Coolant Loop Cold Plates and Rails. The cold plates and rails in the primary coolant loop are arranged in three groups: upstream electronics, aft equipment bay, and batteries.

Coolant from the recirculation assembly flows into parallel paths that serve the upstream electronics cold plate group. In this group, the data storage electronics assembly (DSEA) is cooled by cold rails; the remainder of the electronics, by cold plates. The cold plates are in the pressurized and unpressurized areas of the vehicle. The flow rates through the parallel paths are controlled by flow restrictors, installed downstream of the cold plate group. The first upstream electronics flow path cools the suit circuit heat exchanger. The second flow path cools five cold plates mounted on the pressurized side of the equipment tunnel back wall. The third path serves the integrally cooled IMU and the rate gyro assembly (RGA) cold plate, both located in the unpressurized area (on the navigation base). The fourth path cools the abort sensor assembly (ASA) and pulse torque assembly (PTA) cold plates. All the plates for the fourth path are in the unpressurized area above the cabin; the ASA is on the navigation base of the alignment optical telescope (AOT). The fifth path serves the tracking light electronics (TLE), gimbal angle sequencing transformation assembly (GASTA), lighting control assembly (LCA), and DSEA plates: one in the unpressurized area in front of the cabin, a second one in the control and display panel area, a third one below the cabin floor, and another one on the left wall of the cabin.

The aft equipment bay is cooled by eight cold rails; the flow is in parallel. The batteries are cooled by cold rails. The ascent batteries are in the center section of the aft equipment bay. For the modified LM, the descent batteries are on the -Z outrigger bulkhead of the descent stage. During the descent phase, the coolant flow is split between the descent batteries and the ascent batteries; the ascent batteries are not used during this time. When the stages are separated, quick-disconnects break the coolant lines and seal the ends; all coolant then flows through the ascent battery cold rails.

3-7.5.2. Secondary Coolant Loop Cold Plates and Rails. The secondary coolant loop is for emergency use. Only cold plates and cold rails that have two independent passages (one for the primary loop and one for the secondary loop) are served by this loop.

In the upstream electronics area, the secondary coolant flow is split between three cold plates (RGA, ASA, and TLE) in parallel. The flow rate is controlled by a flow restrictor downstream of the TLE and RGA. After these three plates, the secondary loop cools the ascent battery cold rails and the aft equipment bay cold rails in a series parallel arrangement. The coolant first flows through three ascent battery cold rails in parallel, then through eight aft equipment bay cold rails in parallel.

3-8. COMMUNICATIONS SUBSYSTEM. (See figures 3-8.1 and 3-8.2.)

The Communications Subsystem (CS) provides in-flight and lunar surface communications links between the LM and the CSM, between the LM and Manned Space Flight Network (MSFN), and between the LM and the extravehicular astronaut (EVA). When both astronauts are outside the vehicle, the vehicle relays communications between the astronauts and MSFN. The CS consists of S-band and VHF equipment. The general location of the communications equipment is shown in figure 3-8.3. The mode, band, and purpose of the communications links are given in table 3-8.1. The S-band communications capabilities are given in table 3-8.2.

3-8.1. IN-FLIGHT COMMUNICATIONS. (See figure 3-8.4.)

In-flight, when the vehicle is separated from the CSM and is on the earth side of the moon, the CS provides S-band communications with MSFN and VHF communications with the CSM. When the vehicle and the CSM are on the far side of the moon, VHF is used for communications between them.

3-8.1.1. Earth Side (LM) - MSFN). In-flight S-band communications between the vehicle and MSFN include voice, digital uplink signals and ranging code signals from MSFN. The vehicle S-band equipment transmits voice, acts as transponder to the ranging code signals, transmits biomedical and systems telemetry data, and provides a voice backup capability and an emergency key capability.

S-band voice is the primary means of communication between MSFN and the vehicle. Backup voice communication from MSFN is possible, using the digital uplink assembly, but this unit is normally used by MSFN to update the LM guidance computer. In response to ranging code signals sent to the vehicle, the S-band equipment supplies MSFN with a return ranging code signal that enables MSFN to track, and determine the range of the vehicle. The vehicle transmits biomedical data pertinent to astronaut heartbeat so that MSFN can monitor and record the physical condition of the astronauts. The vehicle also transmits systems telemetry data for MSFN evaluation; voice, using redundant S-band equipment; and, in case there is no vehicle voice capability, provides an emergency key signal so that the astronauts can transmit Morse code to MSFN.

3-8.1.2. Earth Side (LM - CSM). In-flight VHF communications between the LM and the CSM include voice, backup voice, and tracking and ranging signals. Normal LM-CSM voice communications use VHF channel A simplex. Backup voice communication is accomplished with VHF channel B simplex. VHF ranging, initiated by the CSM, used VHF channels A and B duplex.

3-8.1.3. Far Side (LM - CSM). When the LM and the CSM are behind the moon, contact with MSFN is not possible. VHF channel A is used for simplex LM-CSM voice communications. VHF channel B is used as a one-way data link to transmit telemetry signals from the vehicle, to be recorded and stored by the CSM. When the CSM establishes S-band contact with MSFN, the stored data are transmitted by the CSM at 32 times the recording speed.

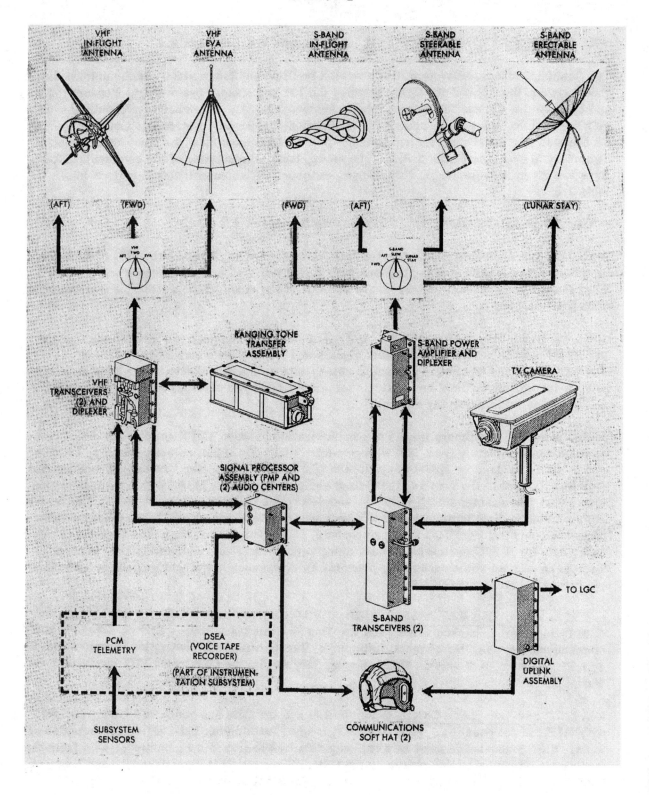

Figure 3-8.1. Communications Subsystem - Flow Diagram

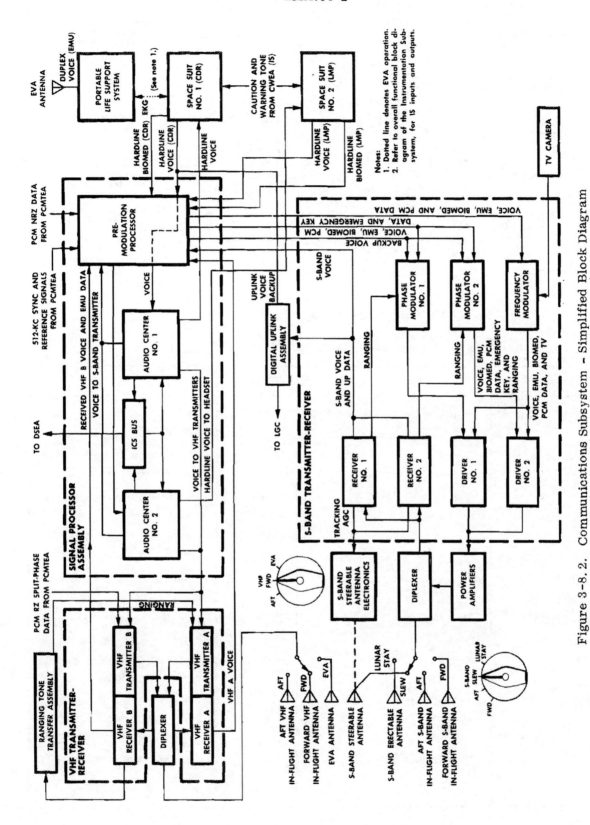

Figure 3-8.2. Communications Subsystem – Simplified Block Diagram

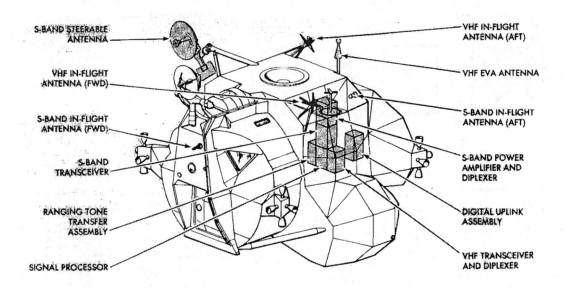

Figure 3-8.3. Communications Subsystem - Equipment Location

Table 3-8.1. Communications Link, Mode, Band, and Purpose

Link	Mode	Band	Purpose
MSFN-LM-MSFN	Pseudorandom noise (PRN)	S-band	Ranging and tracking by MSFN
LM-MSFN	Voice	S-band	In-flight and lunar surface communications
LM-CSM	Voice	VHF simplex	In-flight communications
CSM-LM-MSFN	Voice	VHF and S-band	Conference (with LM as relay)
LM-CSM	Low-bit-rate telemetry	VHF (one way)	CSM records and retransmits to earth
CSM-LM-CSM	Ranging	VHF duplex	Ranging and tracking by CSM
MSFN-LM	Voice	S-band	In-flight and lunar surface communications
MSFN-LM	Uplink data or uplink voice backup	S-band	Update LM guidance computer, or voice backup for in-flight communications
LM-MSFN	Television	S-band	Provides lunar televised data

1 November 1969

Table 3-8.1. Communications Link, Mode, Band, and Purpose (cont)

Link	Mode	Band	Purpose
LM-MSFN	Biomed-PCM telemetry	S-band	Transmission of biomedical and vehicle status data
LM-MSFN-CSM	Voice	S-band	Conference (with earth as relay)
EVA-LM-EVA	Voice and data; voice	VHF duplex	EVA direct communication
EVA-LM-MSFN	Voice and data	VHF, S-band	Conference (with LM as relay)
CSM-MSFN-LM-EVA	Voice and data	S-band, VHF	Conference (via MSFN-LM relay)

Table 3-8.2. S-Band Communications Capabilities

Information	Frequency or Rate	Subcarrier Modulation	Subcarrier Frequency	RF Carrier Modulation
UPLINK: 2101.8 MHz				
Voice	300 to 3000 Hz	FM	30 kHz	PM
Voice backup	300 to 3000 Hz	FM	70 kHz	PM
PRN ranging code	990.6 kilobits/second			PM
Uplink data	1.0 kilobits/second	FM	70 kHz	PM
DOWNLINK: 2282.5 MHz				
Voice	300 to 3,000 Hz	FM	1.25 MHz	PM or FM
Biomed	14.5-kHz subcarrier	FM	1.25 MHz	PM or FM
Extravehicular mobility unit (EMU)	3.9-, 5.4-, 7.35-, and 10.5-kHz subcarriers	FM	1.25 MHz	PM or FM
Voice	300 to 3000 Hz	None	None	Direct PM baseband modulation
Biomed	14.5-kHz subcarrier	None	None	Direct PM baseband modulation

Table 3-8.2.   S-Band Communications Capabilities (cont)

Information	Frequency or Rate	Subcarrier Modulation	Subcarrier Frequency	RF Carrier Modulation
DOWNLINK: 2282.5 MHz				
Extravehicular mobility unit	3.9-, 5.4-, 7.35-, and 10.5- kHz subcarriers	None	None	Direct PM baseband modulation
Voice backup	300 to 3000 Hz	None	None	Direct PM baseband modulation
PRN ranging code (turnaround)	990.6 kilobits/ second			PM
Emergency keying	Morse code	AM	512 kHz	PM
Pulse-code-modulation (PCM) nonreturn-to-zero (NRZ) data	High bit rate: 51.2 kilobits/ second or Low bit rate: 1.6 kilobits/ second	Phase shift keying (PSK)	1.024 MHz	PM or FM
TV	10 to 500 Hz			FM baseband modulation

3-8.2.   LUNAR SURFACE COMMUNICATIONS.   (See figure 3-8.5.)

When the LM is on the lunar surface, the CS provides S-band communications with MSFN and VHF communications with the EVA   The vehicle relays VHF signals to MSFN, using the S-band.

Communications with the CSM may be accomplished by using MSFN as a relay.   LM-MSFN S-band capabilities are the same as in-flight capabilities, except that, in addition, TV may be transmitted from the lunar surface in an FM mode.

3-8.3.   FUNCTIONAL DESCRIPTION.

Each astronaut has his own audio center.   The audio centers have audio amplifiers and switches that are used to route signals between the astronauts, and between the vehicle and MSFN or the CSM.   The centers are redundant in that each one can be used by either astronaut, or both astronauts can use either audio center if necessary.

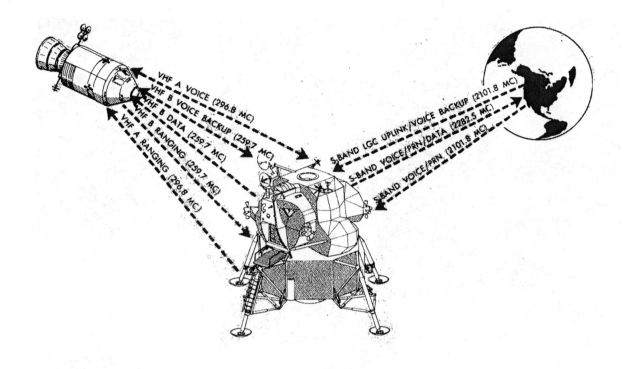

Figure 3-8.4. In-Flight Communications

In the transmission mode, the output of the audio centers goes to the VHF transceivers, or to the premodulation processor (PMP), or to the data storage electronics assembly in the Instrumentation Subsystem (IS). If an audio center output is routed to the VHF transmitter, the transmission is through the diplexer to the selected VHF antenna. If an audio center output is routed to the premodulation processor and then to the S-band transceivers, the transmitter output is applied to the diplexer or to the S-band power amplifier, depending on power output requirements. The output from the transmitter or the power amplifier goes through the diplexer to the selected S-band antenna. If an audio center output is routed to the data storage electronics assembly, the voice transmission is recorded.

The inputs to the S-band transceivers are from the premodulation processor or the television camera. The outputs from the premodulation processor (to be transmitted by S-band transmitters) are processed voice, and PCM, EMU, and biomed data. For television transmission, the S-band power amplifier is used. In normal flight, the steerable antenna is used. On the lunar surface, the S-band erectable antenna is used for normal communications. The S-band omni antennas are used in any one of a number of backup modes.

External RF inputs to the S-band equipment are MSFN voice, either uplink data or an uplink backup voice signal, and ranging. Received MSFN voice is routed through the premodulation processor to the audio centers. Received uplink data signals are routed to the digital uplink assembly, to be decoded and sent to the LM guidance computer. MSFN backup voice is routed to the digital uplink assembly, where it is decoded and then sent to the Commander's microphone amplifier input.

Figure 3-8.5. Lunar Surface Communications

3-8.4. S-BAND TRANSCEIVER.

The S-band transceiver assembly provides deep-space communications between the LM and MSFN. S-band communications consist of voice and pseudorandom noise ranging transmission from MSFN to the vehicle and voice, pseudorandom noise ranging turnaround, biomed, and subsystem data transmission from the vehicle to MSFN. The assembly consists of two identical phase-locked receivers, two phase modulators with driver and multiplier chains, and a frequency modulator. The receivers and phase modulators provide the ranging, voice, emergency-keying, and telemetry transmit-receive functions. The frequency modulator is primarily for video transmission, but accomodates pulse-code-modulation telemetry (subsystem data), biomed, and voice transmission. The frequency modulator provides limited backup for both phase modulators. The operating frequencies of the S-band equipment are 2282.5 MHz (transmit) and 2101.8 MHz (receive).

3-8.5. S-BAND POWER AMPLIFIER.

The S-band power amplifier amplifies the S-band transmitter output when additional transmitted power is required. This assembly consists of two amplitrons, an input and an output isolator (ferrite circulators), and two power supplies, all mounted on a common chassis. The RF circuit is a series interconnection of the isolators and amplitrons. The amplitrons (which are characteristic of saturated, rather than linear, amplifiers) have broad bandwidth, high efficiency, high peak and average power output, but relatively low gain. The isolators protect both amplitrons and both S-band transmitter driver and multiplier chains. The isolators exhibit a minimum isolation of 20 db and a maximum insertion loss of 0.6 db. Each amplitron has its own power supply. One amplitron is designated primary; the other, secondary. Only one amplitron can be activated at a time. When neither amplitron is selected (feedthrough mode), there is a feedthrough path through the power amplifier, with maximum insertion loss of 3.2 db.

3-8.6. VHF TRANSCEIVER.

The VHF transceiver assembly provides voice communications between the LM and the CSM and, during blackout of transmission to MSFN, low-bit-rate telemetry transmission from the vehicle to the CSM, and ranging on the vehicle by the CSM. When the vehicle mission profile includes extravehicular activity, this equipment also provides EVA-LM voice communications, and reception of EVA biomed and suit data for transmission to MSFN over the S-band. The assembly consists of two solid-state superheterodyne receivers and two transmitters. One transmitter-receiver combination provides a 296.8-MHz channel (channel A); the other, a 259.7-MHz channel (channel B), for simplex or duplex voice communications. Channel B may also be used to transmit pulse-code-modulation data from the IS at the low bit rate and to receive biomed and suit data from the EVA during EVA-programmed missions.

3-8.7. SIGNAL PROCESSOR ASSEMBLY.

The signal processor assembly is the common acquisition and distribution point for most CS received and transmitted data, except that low-bit-rate, split-phase data are directly coupled to VHF channel B and TV signals are directly coupled to the S-band transmitter. The signal processor assembly processes voice and biomed signals and provides the interface between the RF electronics, data storage electronics assembly, and pulse-code-modulation and timing electronics assembly of the IS.. The signal processor assembly consists of an audio center for each astronaut and a premodulation processor. The signal processor assembly does not handle ranging and uplink data signals. The premodulation processor provides signal modulation, mixing, and switching in accordance with the selected mode of operation. It also permits the vehicle to be used as a relay station between the CSM and MSFN, and, for EVA-programmed missions, between the EVA and MSFN. The audio centers are identical. They provide individual selection, isolation, and amplification of audio signals received by the CS receivers and which are to be transmitted by the CS transmitters. Each audio center contains a microphone amplifier, headset amplifier, voice-operated relay (VOX) circuit, diode switches, volume control circuits, and isolation pads. The VOX circuit controls the microphone amplifier by activating it only when required for voice transmission. Audio signals are routed to and from the VHF A, VHF B, and S-band equipments and the intercom bus via the audio centers. The intercom bus, common to both audio centers, provides hardline communications between the astronauts. Voice signals to be recorded by the data storage electronics assembly are taken from the intercom bus.

3-8.8. DIGITAL UPLINK ASSEMBLY.

The digital uplink assembly decodes S-band uplink commands from MSFN and routes the data to the LM guidance computer. The digital uplink assembly provides a verification signal to the IS for transmission to MSFN, to indicate that the uplink messages have been received and properly decoded by the digital uplink assembly. The LM guidance computer also routes a no-go signal to the IS for transmission to MSFN whenever the computer receives an incorrect message. The uplink commands addressed to the vehicle parallel those inputs available to the LM guidance computer via the display and keyboard. The digital uplink assembly also provides a voice backup capability if the received S-band audio circuits in the premodulation processor fail.

3-8.9. RANGING TONE TRANSFER ASSEMBLY.

The ranging tone transfer assembly operates with VHF receiver B and VHF transmitter A to provide a transponder function for CSM-LM VHF ranging. The ranging tone transfer assembly receives VHF ranging tone inputs from VHF receiver B and produces ranging tone outputs to key VHF transmitter A.

The VHF ranging tone input consists of two acquisition tone signals and one track tone signal. Accurate ranging is accomplished when the track tone signal from the CSM is received and retransmitted from the vehicle.

3-8-10. S-BAND STEERABLE ANTENNA.

The S-band steerable antenna is a 26-inch-diameter parabolic reflector with a point source feed that consists of a pair of cross-sleeved dipoles over a ground plane. The prime purpose of this antenna is to provide deep space voice and telemetry communications and deep-space tracking and ranging. This antenna provides 174° azimuth coverage and 330° elevation coverage. The antenna can be operated manually or automatically. The manual mode is used for initial positioning of the antenna to orient it within ±12.5° (capture angle) of the line-of-sight signal received from the MSFN station. Once the antenna is positioned within the capture angle, it can operate in the automatic mode.

3-8-11. S-BAND IN-FLIGHT ANTENNAS.

The two S-band in-flight antennas are omnidirectional; one is forward and one is aft on the LM. The antennas are right-hand circularly polarized radiators that collectively cover 90% of the sphere at -3db or better. They operate at 2282.5 MHz (transmit) and 2101.8 MHz (receive). These antennas are the primary S-band antennas for the vehicle during flight.

3-8.12. S-BAND ERECTABLE ANTENNA.

The S-band erectable antenna is stowed in the descent stage. When erected on the lunar surface, it is a 10-foot-diameter parabolic reflector, which is used as a reflector with a telescopic feed system.

3-8.13. VHF IN-FLIGHT ANTENNAS.

The two VHF in-flight antennas are omnidirectional, right-hand circularly polarized antennas that operate in the 259.7- to 296.8-MHz range.

3.8-14. VHF EVA ANTENNA.

The VHF EVA antenna is an omnidirectional conical antenna, which is used for LM-EVA communications when the vehicle is on the lunar surface. It is mounted on the vehicle, and unstowed by an astronaut in the vehicle after landing.

3-9. EXPLOSIVE DEVICES SUBSYSTEM.

The Explosive Devices Subsystem (EDS) is used to perform propellant tank pressurization, landing gear deployment, descent propellant tank venting, and ascent and descent stage separation. (See figure 3-9.1). The explosive devices are electrochemical devices, which are operated by the astronauts.

The LM has two types of explosive devices: detonator cartridges containing high-explosive charges of high yield and pressure cartridges containing propellant charges of relatively low yield. An electrical signal, controlled by the astronauts, triggers an initiator that fires the cartridges. The general location of all the explosive devices is shown in figure 3-9.2.

3-9.1. LANDING GEAR DEPLOYMENT.

Each landing gear assembly is retained in the stowed position by an uplock assembly. The uplock assembly contains two detonator cartridges. While the vehicle is docked with the CM in lunar orbit, both detonator cartridges in each uplock assembly are fired to deploy the landing gear. A landing gear deployment talkback on the control panel turns gray when all four landing gear assemblies have been deployed.

3-9.2. REACTION CONTROL SUBSYSTEM PROPELLANT TANK PRESSURIZATION.

The Reaction Control Subsystem (RCS) fuel and oxidizer tanks are pressurized immediately before landing gear deployment. Two cartridges, which open dual, parallel helium isolation valves, are fired to pressurize the tanks. The RCS can then be operated to separate the LM from the CM.

3-9.3. DESCENT PROPELLANT TANK PRESSURIZATION.

Before starting the descent engine, the descent propellant tanks must be pressurized. Compatibility valve cartridges are fired to open the valves. Cartridges are then fired to open the ambient helium isolation valve. After the descent engine is started, the cryogenic helium flows freely to the descent engine fuel and oxidizer tanks, pressurizing them.

3-9.4. DESCENT PROPELLANT TANK VENTING.

After lunar landing, two explosive vent valves are open to accomplish planned depressurization of the descent propellant tanks. This protects the astronauts, when outside the vehicle, against untimely venting of the tanks through the relief valve assemblies.

3-9.5. ASCENT PROPELLANT TANK PRESSURIZATION.

Before initial start of the ascent engine, the ascent propellant tanks must be pressurized. To accomplish this, explosive valve cartridges, which simultaneously open helium isolation valves and fuel and oxidizer compatibility valves, are fired. This permits helium to pressurize the ascent fuel and oxidizer tanks.

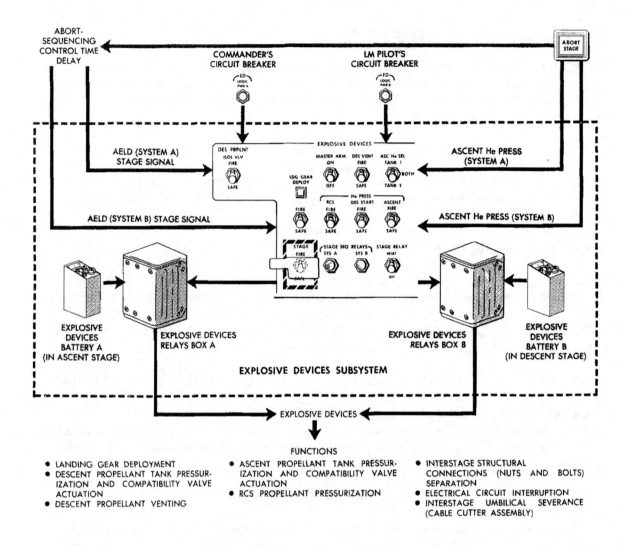

Figure 3-9.1. Explosive Devices Subsystem - Flow Diagram

3-9.6. STAGE SEPARATION.

The ascent and descent stages are separated just before lunar lift-off or if necessary, in the event of mission abort. Control switches are set to initiate a controlled sequence of stage separation. First, all signal and electrical power between the two stages is terminated by explosive circuit interrupters. Next, explosive nuts and bolts joining the stages are ignited. Finally, an explosive guillotine (cable cutter assembly) servers all wires, cables, and water lines connected between the stages. With stage separation completed, operation of its engine can propel the ascent stage into lunar orbit for rendezvous with the CM.

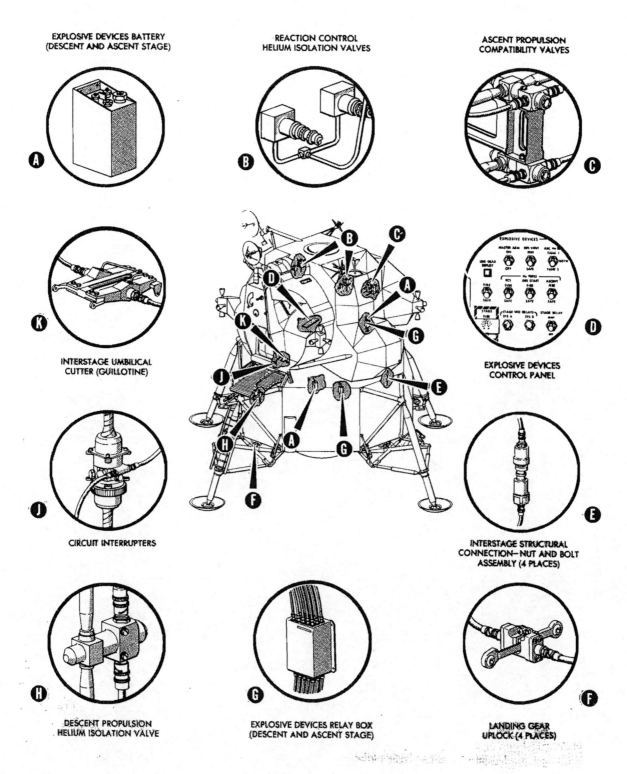

Figure 3-9.2. Explosive Devices Subsystem - Equipment Location

3-10. INSTRUMENTATION SUBSYSTEM. (See figure 3-10.1.)

The Instrumentation Subsystem (IS) monitors the LM subsystems during manned missions, performs in-flight checkout, prepares LM status data for transmission to MSFN, provides timing frequencies and correlated data for LM subsystems, and stores voice and time-correlation data. The IS consists of subsystem sensors, a signal-conditioning electronics assembly (SCEA), pulse-code-modulation and timing electronics assembly (PCMTEA), caution and warning electronics assembly (CWEA), and data storage electronics assembly (DSEA). Each IS assembly receives operating power from the Electrical Power Subsystem (EPS) through associated circuit breakers on the Commander and LM Pilot circuit breaker panels.

3-10.1. SUBSYSTEM SENSORS.

The subsystem sensors are located throughout the LM subsystems and structure. These sensors continuously monitor the status of the various subsystems parameters, such as temperature, quantity, frequency, valve position, pressure, switch position, voltage, and current. The sensors convert the sensed physical parameters into electrical output signals (digital or analog) compatible with the requirements of the SCEA or PCMTEA. Output signals that require conditioning, are routed to the SCEA. Output signals that do not require conditioning are routed to the PCMTEA.

In essence, the modified LM sensing and data-acquisition requirements are identical with that of the LM. The modified LM requires installation of additional transducers (sensors) and relocation of a few others to satisfy additional measurement requirements.

3-10.2. SIGNAL-CONDITIONING ELECTRONICS ASSEMBLY.

The SCEA converts all unconditioned sensor signals and events to proper voltage levels required by the PCMTEA, CWEA, and displays. The unconditioned signals are fed through amplifiers, attenuators, ac-to-dc converters, analog and discrete isolating buffers, frequency-to-dc converters, resistance-to-dc converters, and phase-sensitive demodulators to provide the proper output voltage. These seven basic SCEA subassemblies are in each electronic replaceable assembly (ERA-1 and ERA-2) that makes up the SCEA. Each ERA has capacity for 22 plug-in assemblies (conditioning units) and provides the interface connections between other LM vehicle subsystems and the IS. These subassemblies provide conditioned digital and analog data. Digital data (discrete event functions) outputs appear as a 4- to 6-volt d-c level for logic 1 or "on"; as a 0- to 0.5-volt d-c level, for logic 0 or "off". Discrete signals to be monitored by displays are in the form of solid-state switch closures. Analog data varies from 0 to 5 volts dc. Each ERA contains circuits that route signals to IS assemblies.

Because of the additional measurement requirements for the modified LM, three plug-in subassemblies (conditioning units) were added. A resistance-to-dc-converter (506-3) was added to both ERA-1 and ERA-2.

3-10.3. PULSE-CODE-MODULATION AND TIMING ELECTRONICS ASSEMBLY.

The PCMTEA prepares subsystem status signals for transmission and provides timing frequencies and mission-elapsed-time data to LM subsystems. It consists of pulse-code-modulation (PCM) equipment and timing electronics (TE) equipment. The PCMTEA receives a 1,024-kpps synchronizing signal and three types of data: high-level analog, parallel digital, and serial digital. Some parallel digital and high-level analog inputs are supplied directly to the PCMTEA from the sensors; others, are routed through the SCEA. The data inputs routed to the PCMTEA are changed to serial digital form for transmission to MSFN at one of two bit rates: 1,600 bits per second (low rate) or 51,200 bits per second (high rate). Subcarrier frequencies, time reference signals, and sync pulses are generated in the PCMTEA and supplied to other subsystems that require them for proper operation. The 1,024-kpps sync signal from the LM guidance computer (LGC) of the Guidance, Navigation and Control Subsystem (GN&CS) provides synchronization for the PCMTEA and those subsystems that use outputs from the TE equipment of the PCMTEA.

3-10.3.1. <u>Pulse-Code-Modulation Equipment</u>. The PCM equipment processes subsystem data for transmission. This equipment consists of analog multiplexer gate drivers, high-level analog multiplexer and high-speed gates, an analog-to-digital converter (coder), a calibrator, programmer, digital multiplexer, an output register and data transfer buffers, and a power supply.

Primary control of all units of the PCM equipment is established by the programmer. Using submultiples of the internal or external 1,024-kpps signals, the programmer provides the basic timing for data sampling and processing. It generates signals for bit, word, prime frame, and subframe timing intervals. External format select signals cause the programmer to generate commands for a fixed, predetermined program of analog and digital data sampling and to sequentially control the sampling operation. The programmer generates synchronization and format identification word patterns and inserts this information into the output serial data stream. It also generates the command and timing signals for synchronizing other LMMP vehicle equipment with the PCMTEA.

The analog multiplexer drivers receive the timing and command signals from the programmer and route the signals to appropriate circuits in the high-level analog multiplexer gates. The analog voltage inputs to these gates are selected at the programmed sampling rate and applied to the high-speed gates. The high-level pulse-amplitude-modulated (PAM) outputs from the high-speed gates are routed to the coder, which produces digitized words that represent the input analog data and supplies the words to the digital multiplexer.

The digital multiplexer gates receive digitized-data signals from the coder and pure digital information from other LM equipment. Gating command signals from the programmer control sequential operation of the digital multiplexer gates according to a fixed program, producing multiplexed data that consists of parallel binary words. The parallel binary words are gated to the output register and data transfer buffers (output buffers). The output register converts the parallel data to a serial RZ and NRZ-C output. It also receives, from the LGC and the abort guidance section (AGS) of the GN&CS, serial digital data, which are inserted into the serial output of the PCMTEA.

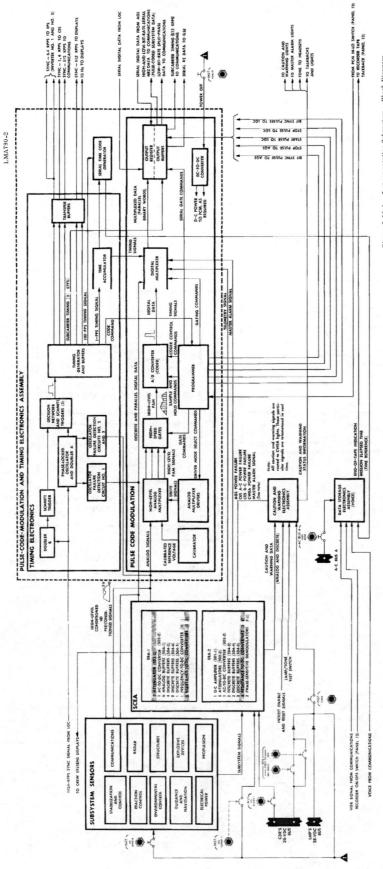

Figure 3-10.1. Instrumentation Subsystem – Block Diagram

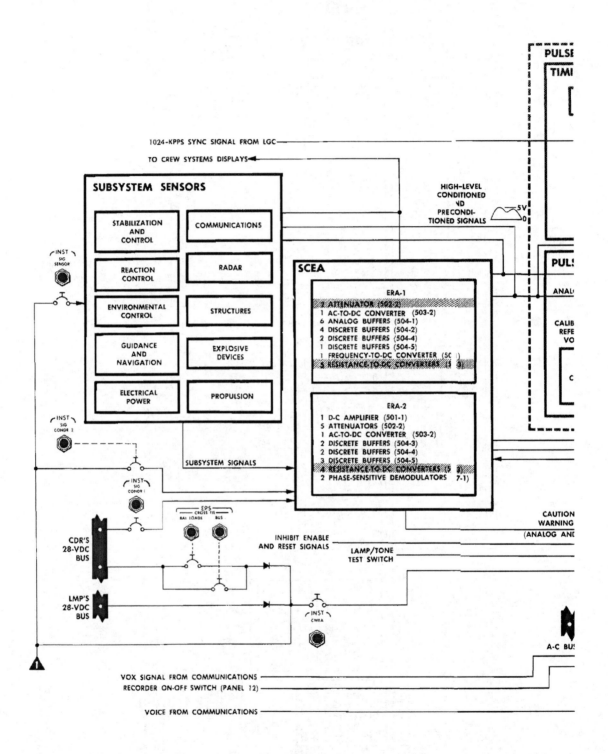

1024-KPPS SYNC SIGNAL FROM LGC

TO CREW SYSTEMS DISPLAYS

PULSE

TIMI

HIGH–LEVEL
CONDITIONED
ND
PRECONDI-
TIONED SIGNALS

5V
0

SUBSYSTEM SENSORS

STABILIZATION AND CONTROL	COMMUNICATIONS
REACTION CONTROL	RADAR
ENVIRONMENTAL CONTROL	STRUCTURES
GUIDANCE AND NAVIGATION	EXPLOSIVE DEVICES
ELECTRICAL POWER	PROPULSION

INST
SIG
SENSOR

INST
SIG
CONDR 2

SCEA

PULS

ANALO

CALIB
REFE
VO

C

ERA-1

2 ATTENUATOR (502-2)
1 AC-TO-DC CONVERTER (503-2)
6 ANALOG BUFFERS (504-1)
4 DISCRETE BUFFERS (504-2)
2 DISCRETE BUFFERS (504-4)
1 DISCRETE BUFFERS (504-5)
1 FREQUENCY-TO-DC CONVERTER (5C)
5 RESISTANCE-TO-DC CONVERTERS (5 3)

ERA-2

1 D-C AMPLIFIER (501-1)
5 ATTENUATORS (502-2)
1 AC-TO-DC CONVERTER (503-2)
2 DISCRETE BUFFERS (504-3)
2 DISCRETE BUFFERS (504-4)
3 DISCRETE BUFFERS (504-5)
4 RESISTANCE-TO-DC CONVERTERS (5 8)
2 PHASE-SENSITIVE DEMODULATORS 7-1)

SUBSYSTEM SIGNALS

INST
SIG
CONDR 1

EPS
CROSS TIE
BAT LOADS   BUS

CAUTION
WARNING
(ANALOG AND

INHIBIT ENABLE
AND RESET SIGNALS

LAMP/TONE
TEST SWITCH

CDR'S
28-VDC
BUS

LMP'S
28-VDC
BUS

INST
CWEA

A-C BUS

VOX SIGNAL FROM COMMUNICATIONS
RECORDER ON-OFF SWITCH (PANEL 12)

VOICE FROM COMMUNICATIONS

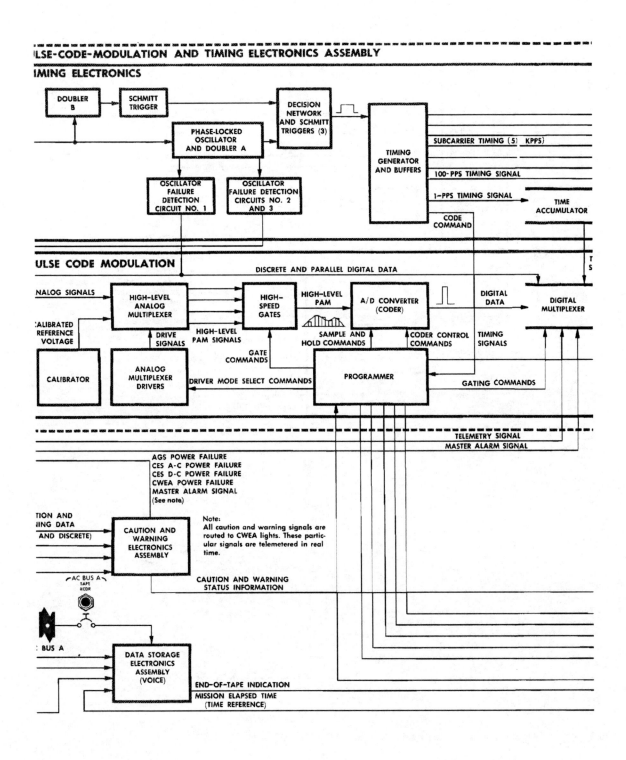

TIMING ELECTRONICS

DOUBLER B

SCHMITT TRIGGER

DECISION NETWORK AND SCHMITT TRIGGERS (3)

PHASE-LOCKED OSCILLATOR AND DOUBLER A

TIMING GENERATOR AND BUFFERS

SUBCARRIER TIMING (5) KPPS)

OSCILLATOR FAILURE DETECTION CIRCUIT NO. 1

OSCILLATOR FAILURE DETECTION CIRCUITS NO. 2 AND 3

100-PPS TIMING SIGNAL

1-PPS TIMING SIGNAL

TIME ACCUMULATOR

CODE COMMAND

PULSE CODE MODULATION

DISCRETE AND PARALLEL DIGITAL DATA

ANALOG SIGNALS

HIGH-LEVEL ANALOG MULTIPLEXER

HIGH-SPEED GATES

HIGH-LEVEL PAM

A/D CONVERTER (CODER)

DIGITAL DATA

DIGITAL MULTIPLEXER

CALIBRATED REFERENCE VOLTAGE

DRIVE SIGNALS

HIGH-LEVEL PAM SIGNALS

SAMPLE AND HOLD COMMANDS

CODER CONTROL COMMANDS

TIMING SIGNALS

CALIBRATOR

ANALOG MULTIPLEXER DRIVERS

GATE COMMANDS

DRIVER MODE SELECT COMMANDS

PROGRAMMER

GATING COMMANDS

TELEMETRY SIGNAL

MASTER ALARM SIGNAL

AGS POWER FAILURE
CES A-C POWER FAILURE
CES D-C POWER FAILURE
CWEA POWER FAILURE
MASTER ALARM SIGNAL
(See note)

TION AND ING DATA
AND DISCRETE)

CAUTION AND WARNING ELECTRONICS ASSEMBLY

Note:
All caution and warning signals are routed to CWEA lights. These particular signals are telemetered in real time.

AC BUS A
TAPE RCDR

CAUTION AND WARNING STATUS INFORMATION

BUS A

DATA STORAGE ELECTRONICS ASSEMBLY (VOICE)

END-OF-TAPE INDICATION

MISSION ELAPSED TIME (TIME REFERENCE)

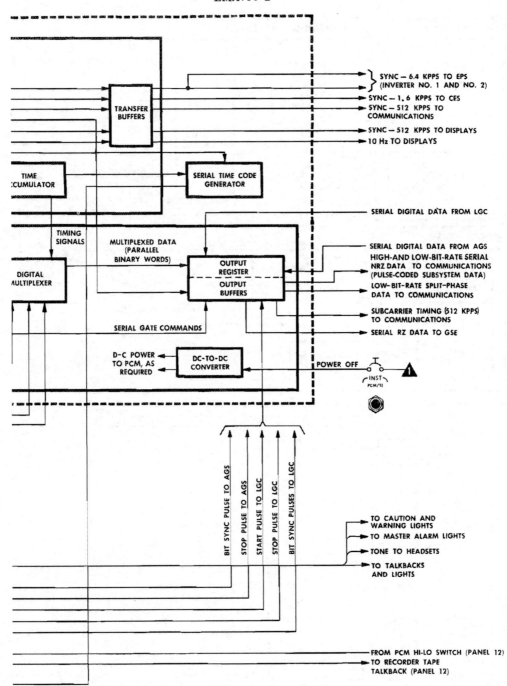

Figure 3-10.1. Instrumentation Subsystem - Block Diagram

3-10.3.2. Timing Electronics Equipment. The TE equipment provides timing and synchronizing signals to the PCM equipment and other LM subsystems. The TE equipment consists of a phase-locked oscillator and doublers, a decision network, a timing generator, time accumulator, serial time code generator, and three oscillator failure-detection circuits. The TE equipment is redundant in critical areas. A 1,024-kpps input from the LGC forces the phase-locked oscillator into synchronism. If the input from the LGC fails, the phase-locked oscillator continues to furnish the necessary timing signals without loss of data. The TE equipment divides the 1,024-kpps signal into a number of selected subharmonic frequencies, which are routed to the PCM equipment and other LM vehicle devices.

3-10.4. CAUTION AND WARNING ELECTRONICS ASSEMBLY. (See figure 3-10.2.)

The CWEA provides the astronauts and MSFN with a continuous rapid check of LM status during manned missions. By continuously monitoring the data supplied by the SCEA, malfunctions are immediately detected. When a malfunction is detected, the CWEA provides signals that light caution lights, warning lights, component caution lights, and MASTER ALARM pushbutton/lights. A malfunction also activates tone in the astronaut headsets and supplies the PCMTEA with a telemetry signal.

The CWEA compares analog signals from the SCEA with preselected internally generated limits supplied by the caution and warning power supply as reference voltage. In addition to the analog inputs, the CWEA receives discrete on-off and contact closure signals. All inputs are applied to detectors (analog and discrete) in the CWEA. The detected signal is routed through logic circuitry, enabling the necessary relay contacts for the caution or warning lights. Simultaneously, the signal is routed to a master flip-flop that energizes a master relay driver, enabling relay contacts. These relay contacts route the signal to light the MASTER ALARM pushbutton/lights and provide the 3-kHz tone to the astronaut headsets. Pressing either MASTER ALARM pushbutton/light extinguishes both lights and terminates the tone, but has no effect on the caution or warning lights. The MASTER ALARM pushbutton/lights are not resettable when the C/W PWR caution light goes on.

The CWEA channel detector logic also receives automatic reset signals from the power supply assembly, and manual caution and manual warning reset signals from vehicle subsystems. The thrust chamber assembly (TCA) logic accepts command input signals that correspond to TCA operation. Automatic reset signals from the power supply assembly and reset signals from LM subsystems are accepted by the TCA logic. TCA outputs are routed simultaneously to a warning relay driver and a talkback relay driver, which operate the RCS TCA warning light and a particular quad talkback, depending on the quadrant in which the TCA (system A or B) failed. The master flip-flop monitors the output of all caution and warning relay drivers. When one of these relay drivers is energized, the master flip-flop energizes master relay drivers. These drivers operate relays whose contacts provide a path to actuate an electronic switch and a telemetry channel. When the electronic switch contacts close, a signal is routed to light both MASTER ALARM pushbutton/lights. Resetting the MASTER ALARM pushbutton/lights does not affect a caution, warning, telemetry, component caution lights, or talkbacks. These malfunction indications remain until the cause is eliminated or until the appropriate

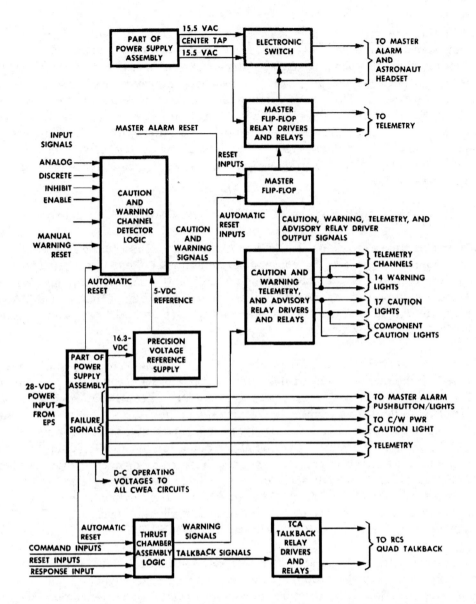

Figure 3-10.2.   Caution and Warning Electronics Assembly -
Functional Block Diagram

reset or inhibit signal is applied.   The telemetry channel provides MSFN with the malfunction indication.   The power supply assembly not only provides operating voltages for
the CWEA; it has automatic reset and failure-detection circuits.   The automatic reset
circuits operate when power is turned on; they reset the master flip-flop and all resettable
CWEA logic circuits.   Power supply failure is sensed by the failure-detection circuit,
which operates relays that cause the MASTER ALARM pushbutton/lights and the C/W PWR
caution light to go on; the relays also close a telemetry channel.   The 16.3-volt d-c output
of the power supply assembly provides a regulated voltage to a precision voltage reference
supply.   This supply provides a precise zener voltage reference for comparing input
signal voltages in the CWEA channel detector logic circuits.

3-10.5. DATA STORAGE ELECTRONICS ASSEMBLY.

The DSEA is a single-speed, four-track, magnetic-tape recorder that stores voice and time-correlation data (TCD) (mission elapsed time). A maximum of 10 hours of recording time is provided (2.5 hours on each track) by driving the tape in one direction over the record head and, on completion of a pass, switching to the next track and reversing tape direction. The tape is supplied in a magazine. Once the magazine is properly placed in the DSEA and the control logic is placed in track No. 1 forward condition (reset), the DSEA is operated with a switch in conjunction with the VOX trigger signal supplied by the signal-processing assembly of the Communications Subsystem.

3-11. LIGHTING.

Lighting is used for LM and CSM tracking and docking maneuvers.

Lighting is provided by exterior and interior lights and lighting control equipment to aid in the performance of crew visual tasks and lessen astronaut fatigue and interior-exterior glare effects. The exterior lighting enables the astronauts to guide and orient the vehicle visually to the CSM to achieve successful tracking and docking. Interior lighting is divided into seven categories: incandescent annunciators, component caution lights, floodlights, computer condition lights, integral electroluminescent lighting, numeric electroluminescent lighting, and incandescently illuminated pushbuttons.

3-11.1. EXTERIOR LIGHTING.

Exterior lighting includes five docking lights, and a high-intensity tracking light (See figure 3-11-1).

3-11.1.1. Docking Lights. Five docking lights mounted on the exterior of the vehicle provide visual orientation and permit gross attitude determination relative to a line of sight through the CSM windows during rendezvous and docking. For transposition and docking, the docking lights are turned on by a switch located at spacecraft-Lunar Module adapter attachment points. This switch is automatically closed upon deployment of the adapter panels. At completion of the docking maneuver, LM power is turned off and the docking lights go off. The lights are visible, and their color recognizable, at a maximum distance of 1,000 feet.

3-11.1.2. Tracking Light. The tracking light permits visual tracking of the LM by the CSM. A flash tube in the tracking light electronics assembly causes the light, which has a 60° beam spread, to flash at a rate of 60 flashes per minute.

3-11.2. INTERIOR LIGHTING.

Interior lighting consists of integral panel and display lighting, backup floodlighting, and electroluminescent lighting. Electroluminescence is light emitted from a crystalline phosphor (ZnS) placed as a thin layer between two closely spaced electrodes of an electrical capacitor; one of the electrodes must be transparent. The light output varies with voltage. Advantageous characteristics are even illumination, low power consumption, and negligible heat dissipation.

3-11.2.1. Integrally Lighted Components. There are three types of integrally lighted components: panel areas, displays, and caution and warning annunciators. The integrally lighted components use electroluminescent or incandescent devices that are controlled by on-off switches and potentiometer-type dimming controls. All panel placards are integrally lighted by white electroluminescent lamps with overlays. The displays have electroluminescent lamps within their enclosures. The numeric displays show green or white illuminated digits on a nonilluminated background; displays with pointers have a nonilluminated pointer travelling over an illuminated background. The brightness of the electroluminescent displays is varied with dimming controls, which can be bypassed by a related override switch, so that full brightness will be maintained should a dimming control fail.

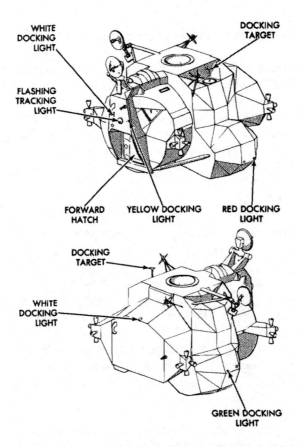

Figure 3-11.1. Ascent Stage Exterior Lighting

3-11.2.2. Lunar Contact Lights. Two lunar contact lights go on when one or more of the three lunar surface sensing probes contact the lunar surface. A probe is mounted beneath three of the landing gear footpads.

3-11.2.3. Floodlighting. Floodlighting is used for general cabin illumination and as a secondary source of illumination for the control and display panels. Floodlighting is provided by the Commander's overhead and forward floodlights, the LM Pilot's overhead and forward floodlights, and recessed floodlights in the bottom of extending side panels. These floodlights provide even distribution of light with minimum reflection.

3-11.2.4. Portable Utility Lights. Two portable utility lights are used, when necessary, to supplement the cabin interior lighting. The lights, when removed from the flight data file container, connect to the overhead utility light panel. Switches provide one-step dimming for light-intensity control.

3-11.2.5. Optical Sight Reticle Light. The crewman's optical alignment sight, used to sight the docking target on the CSM, has a reticle that is illuminated by a 28-volt d-c lamp.

3-11.2.6. <u>Alignment Optical Telescope Lights</u>. A thumbwheel on the computer control and reticle dimmer assembly controls the brightness of the telescope reticle. The lamps edge-light the reticle with incandescent red light.

3-12. CREW PERSONAL EQUIPMENT. (See figure 3-12-1.)

Crew personal equipment (CPE) includes a variety of mission-oriented equipment required for life support and astronaut safety, and accessories related to successful completion of the mission. The equipment ranges from astronaut space suits and docking aids to personal items stored throughout the cabin.

In the modified LM, the following changes have been made to the CPE.

- New MESA with portable equipment pallets and modified sample return equipment table

- Provisions for shirtsleeve operation in cabin

- Liquid waste management system

- Modularized ascent stage stowage provisions

- Third sample return container

The modularized equipment stowage assembly (MESA) and the Apollo lunar scientific equipment package (ALSEP) are stored in the descent stage. This equipment is used for sample and data collecting and scientific experimenting. The resultant data will be used to derive information on the atmosphere and distance between earth and the moon.

The portable life support system (PLSS) interfaces with the Environmental Control Subsystem (ECS), for refills of oxygen and water. The pressure garment assembly or the advanced extravehicular suit interface with the ECS for conditioned oxygen, through oxygen umbilicals, and with the Communications and Instrumentation Subsystems for communications and bioinstrumentation, through the electrical umbilical.

3-12.1. EXTRAVEHICULAR MOBILITY UNIT.

The extravehicular mobility unit (EMU) provides life support in a pressurized or unpressurized cabin, and up to 4 hours of extravehicular life support.

In its extravehicular configuration, the EMU is a closed-circuit pressure vessel that envelops the astronaut. The environment inside the pressure vessel consists of 100% oxygen at a nominal pressure of 3.75 psia. The oxygen is provided at a flow rate of 6 cfm. The extravehicular life support equipment configuration includes the following:

      Liquid cooling garment (LCG)
      Pressure garment assembly (PGA)
      Integrated thermal micrometeoroid garment (ITMG)
      Portable life support system (PLSS)
      Secondary life support system (SLSS)
      Communications carrier
      EMU waste management system
      EMU maintenance kit
      PLSS remote control unit
      Extravehicular visor assembly (EVVA)
      Lunar extravehicular visor assembly (LEVA)
      Biomedical belt

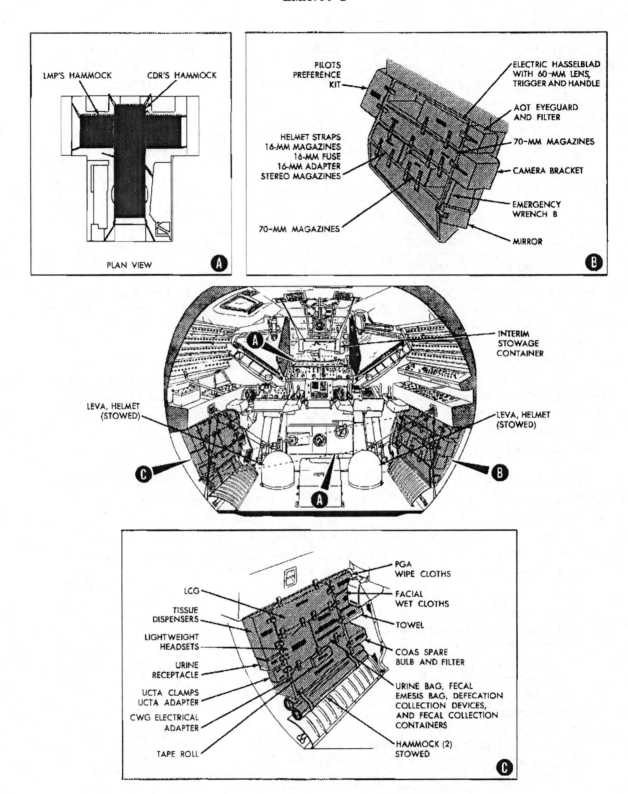

Figure 3-12.1. LM Crew Equipment Arrangement (Sheet 1 of 2)

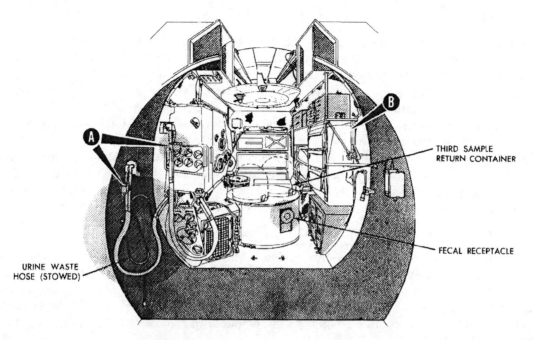

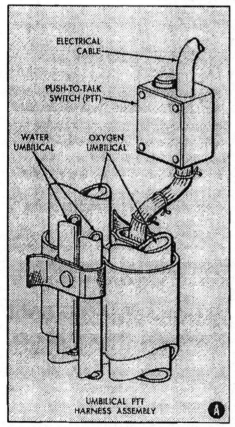

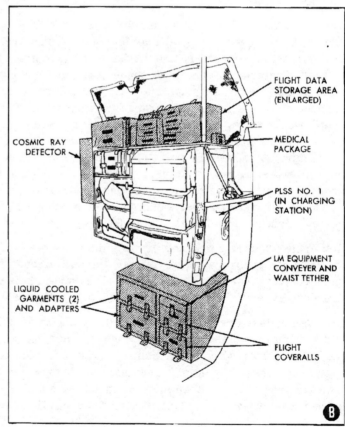

Figure 3-12.1.   LM Crew Equipment Arrangement (Sheet 2 of 2)

3-12.2. LIQUID COOLING GARMENT.

Although shirtsleeve operation is possible in the cabin, the LCG may also be worn. The LCG is worn during all extravehicular activity. It cools the astronaut's body by absorbing body heat and transferring excessive heat to the sublimator in the PLSS. The LCG is a one-piece, long-sleeved, integrated-stocking undergarment of netting material. It consists of an inner liner of Beta cloth, to facilitate donning, and an outer layer of Beta cloth into which a network of Tygon tubing is woven. The tubing does not pass through the stocking area. A double connector for incoming and outgoing water is located on the front of the garment. Cooled water, supplied from the PLSS, is pumped through the tubing. Pockets for bioinstrumentation signal conditioners are located around the waist. A zipper that runs up the front is used for donning and doffing the LCG; and opening at the crotch is used for urinating. Dosimeter pockets and snaps for attaching a biomedical belt are part of the LCG.

3-12.3. PRESSURE GARMENT ASSEMBLY. (See figure 3-12.2.)

The PGA (soft suit) is the basic pressure vessel of the EMU. It can provide a mobile life-support chamber if cabin pressure is lost due to leaks or puncture of the vehicle. The PGA consists of a helmet, torso and limb suit, intravehicular gloves, and various controls and instrumentation to provide the crewman with a controlled environment. The PGA is designed to be worn for 115 hours, in an emergency, at regulated pressure of $3.75\underline{+}0.25$ psig, in conjunction with the LCG.

The torso and limb suit is a flexible pressure garment that encompasses the entire body, except the head and hands. It has four gas connectors, a PGA multiple water receptacle, a PGA electrical connector, and a PGA urine transfer connector for the PLSS/PGA and ECS/PGA interface. The PGA connectors have positive locking devices and can be connected and disconnected without assistance. The gas connectors comprise an oxygen inlet and outlet connector, on each side of the suit front torso. Each oxygen inlet connector has an integral ventilation diverter value. The PGA multiple water receptacle, mounted on the suit torso, serves as the interface between the LCG multiple water connector and PLSS multiple water connector. A protective external cover provides PGA pressure integrity when the LCG multiple water connector is removed from the PGA water receptacle. The PGA electrical connector, provides a communications, instrumentation, and power interface to the PGA. The PGA urine transfer connector on the suit right leg is used to transfer urine from the urine collection transfer assembly (UCTA) to the waste management system.

The urine transfer connector permits dumping the urine collection bag without depressurizing the PGA. A pressure relief valve on the suit sleeve, near the wrist ring, vents the suit in the event of overpressurization. If the valve does not open, it can be manually overridden. A pressure gage on the other sleeve indicates suit pressure.

The helmet is a Lexan (polycarbonate) shell with a bubble-type visor, a vent-pad assembly, and a helmet-attaching ring. The vent-pad assembly permits a constant flow of oxygen over the inner front surface of the helmet. The astronaut can turn his head within the helmet neck-ring area. The helmet does not turn independently of the torso and limb suit. The helmet has provisions on each side for mounting an EVA. When the vehicle is unoccupied, the helmet protective bags are stowed on the cabin floor, at the crew flight stations.

The intravehicular gloves are worn during operations in the cabin. The gloves are secured to the wrist rings of the torso and limb suit with a slide lock; they rotate by means of a

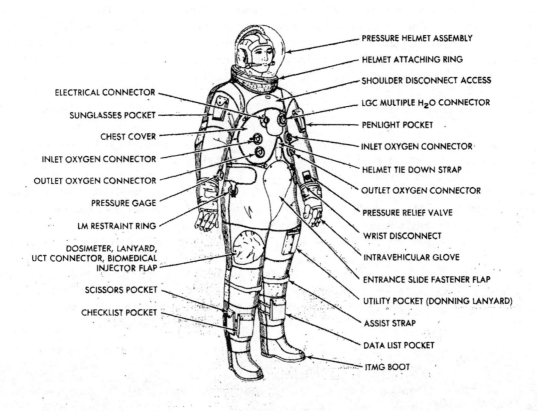

Figure 3-12.2. Pressure Garment Assembly

ball-bearing race. Freedom of rotation, along with convoluted bladders at the wrists and adjustable antiballooning restraints on the knuckle areas, permits manual operations while wearing the gloves.

All PGA controls are accessible to the crewman during intravehicular and extravehicular operations. The PGA controls comprise two ventilation diverter valves, a pressure relief valve with manual override, and a manual purge valve. For intravehicular operations, the ventilation diverter valves are open, dividing the PGA inlet oxygen flow equally between the torso and helmet of the PGA. During extravehicular operation, the ventilation diverter valves are closed and the entire oxygen flow enters the helmet. The pressure relief valve accommodates flow from a failed-open primary oxygen pressure regulator. If the pressure relief valve fails open, it may be manually closed. The purge valve interfaces with the PGA through the PGA oxygen outlet connector. Manual operation of this valve initiates an 8-pound/hour purge flow, providing CO_2 washout and minimum cooling during contingency or emergency operations. A pressure transducer on the right cuff indicates pressure within the PGA. Biomedical instrumentation comprises an EKG (heart) sensor, ZPN (respiration rate) sensor, dc-to-dc converter, and wiring harness. A personal radiation dosimeter (active) is attached to the integrated thermal micrometeoroid garment for continuous accumulative radiation readout. A chronograph wristwatch (elapsed-time indicator) is readily accessible to the crewman for monitoring.

3-12.3.1. Communications Carrier. The communications carrier (cap) is a poly-urethane-foam headpiece with two independent earphones and microphones, which are connected to the suit 21-pin communications electrical connector. The communications carrier is worn with or without the helmet during intravehicular operations. It is worn with the helmet during extravehicular operations.

3-12.4. INTEGRATED THERMAL MICROMETEOROID GARMENT. (See figure 3-12.3.)

The ITMG, worn over the PGA, protects the astronaut from harmful radiation, heat transfer, and micrometeoroid activity. It is a one-piece, form-fitting, multilayered garment that is laced over the PGA and remains with it. The EVA, gloves, and boots are donned separately. From the outer layer in, the ITMG is made of a protective cover, a micro-meteoroid-shielding layer, a thermal-barrier blanket (multiple layers of aluminized mylar), and a protective liner. For extravehicular activity, the PGA gloves are replaced with the extravehicular gloves. The extravehicular gloves are made of the same material as the ITMG to permit handling intensely hot or cold objects outside the cabin and for protection against lunar temperatures. The extravehicular boots (lunar overshoes) are worn over the PGA boots for extravehicular activity. They are made of the same material as the ITMG. The soles have additional insulation for protection against intense tempera-tures.

3-12.5. LUNAR EXTRAVEHICULAR VISOR ASSEMBLY.

The LEVA furnishes visual, thermal, and mechanical protection to the helmet and head. It is composed of a plastic shell, sun blinders, and two visors. The outer (sun) visor is made of polysulfone plastic. The inner protective visor is made of UV-stabilized polycarbonate plastic. The outer visor filters visible light and rejects a significant amount of ultraviolet and infrared rays. The inner visor filters ultraviolet rays and, in combination with the pressure helmet, forms an effective thermal barrier. The two visors, in combination, protect the pressure helmet from micrometeoroid damage and from damage in the event of impact with the lunar surface. A hard shell protects the sun visor when the visor is not used.

The sun visor may be positioned anywhere between "full up" and "full down" if the protective visor is "full down." The force required for moving either visor is 2 to 3 pounds. This force has been determined as necessary to prevent inadvertent movement of either visor from a selected position. An astronaut can attach or detach the LEVA from his helmet without the aid of tools. A latching mechanism allows the lower rim of the LEVA to be tightened and secured around the neck area of the pressure helmet. The mechanism consists of an overcenter latch, which locks on the lower rim, draws the two sides together, and hold them secure. The LEVA/PGA interface collar provides thermal protection for the neck ring.

3-12.6. PORTABLE LIFE SUPPORT SYSTEM.

The PLSS is a self-contained, self-powered, rechargeable environmental control system. In the extravehicular configuration of the EMU, the PLSS is worn on the astronaut's back.

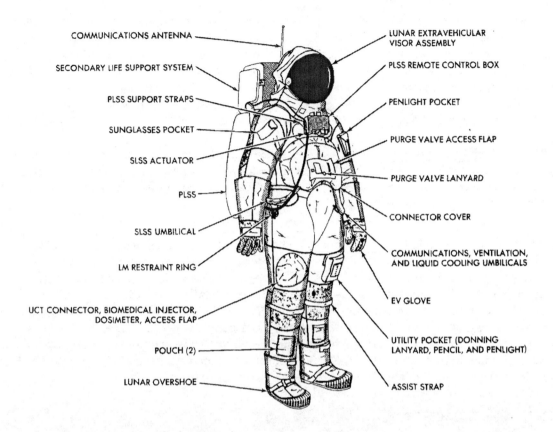

COMMUNICATIONS ANTENNA

SECONDARY LIFE SUPPORT SYSTEM

PLSS SUPPORT STRAPS

SUNGLASSES POCKET

SLSS ACTUATOR

PLSS

SLSS UMBILICAL

LM RESTRAINT RING

UCT CONNECTOR, BIOMEDICAL INJECTOR, DOSIMETER, ACCESS FLAP

POUCH (2)

LUNAR OVERSHOE

LUNAR EXTRAVEHICULAR VISOR ASSEMBLY

PLSS REMOTE CONTROL BOX

PENLIGHT POCKET

PURGE VALVE ACCESS FLAP

PURGE VALVE LANYARD

CONNECTOR COVER

COMMUNICATIONS, VENTILATION, AND LIQUID COOLING UMBILICALS

EV GLOVE

UTILITY POCKET (DONNING LANYARD, PENCIL, AND PENLIGHT)

ASSIST STRAP

Figure 3-12.3.   Integrated Thermal Micrometeoroid Garment

The PLSS supplies pressurized oxygen to the PGA, cleans and cools the expired gas, circulates cool liquid in the LCG through the liquid transport loop, transmits astronaut biomedical data, and functions as a dual VHF transceiver for communication.

The PLSS has a contoured fiberglass shell to fit the back, and a thermal micrometeoroid protective cover. It has three control valves and, on a separate remote control unit, two control switches, a volume control, and a five-position switch for the dual VHF transceiver. The remote control unit is set on the chest.

The PLSS attaches to the astronaut's back, over the ITMG; it is connected by a shoulder harness assembly. When not in use, it is stowed on the floor or in the left-hand mid-section. To don the PLSS, it is first hooked to the overhead attachments in the left-hand midsection ceiling. The astronaut backs against the pack, makes PGA and harness connections, and unhooks the PLSS straps from the overhead attachment.

The PLSS can operate for 4 hours in space environment before oxygen and feedwater must be replenished and the battery replaced. The basic systems and loops of the PLSS are a primary oxygen subsystem, an oxygen ventilation loop, a feedwater loop, liquid transport loop, and an electrical system.

The space suit communicator (SSC) in the PLSS provides primary and secondary duplex voice communication and physiological and environmental telemetry. All EMU data and voice must be relayed through the LM and CM and transmitted to MSFN via S-band. The VHF antenna is permanently mounted on the secondary life support system (SLSS). Two tone generators in the SSC provide audible 3- and 1.5-kHz warning tones to the communications cap receivers. The generators are automatically turned on by high oxygen flow, low vent flow, or low PGA pressure. Both tones are readily distinguishable.

3-12.6.1. <u>PLSS Remote Control Unit</u>. The PLSS remote control unit is a chest-mounted instrumentation and control unit. It has a fan switch, pump switch, SSC mode selector switch, volume control, PLSS oxygen quantity indicator, five status indicators, and an interface for the SLSS actuator.

3.12.6.2 <u>Secondary Life Support System</u>. The secondary life support system (SLSS) is a self contained, independently powered, nonrechargeable emergency oxygen system. It provides 90 minutes of oxygen for lunar surface activities and 30 minutes of oxygen for orbital extravehicular activities. The SLSS is essentially a miniaturized PLSS, but does not contain a communications system. The SLSS supplies pressurized oxygen to the PGA, cleans and cools the expired gases, and supplies cooled water to the LCG from a rechargeable liquid cooling system. In the normal extravehicular configuration, the SLSS is mounted on top of the PLSS; for contingency operation, the SLSS is attached to the PGA front lower torso. A SLSS for each astronaut is stowed in the LM.

3-12.7. UMBILICAL ASSEMBLY.

The umbilical assembly consists of separable flexible hoses and connectors for securing the PGA to the ECS, Communications Subsystem (CS), Instrumentation Subsystem (IS), oxygen, water and electrical umbilicals, to each astronaut.

The oxygen umbilical consists of hoses (1.25-inch inside diameter) with corrosion resistant wire reinforcement. The connectors are of the quick-disconnect type, with a 1.24-inch 90° elbow at the PGA end. Each assembly is made up of two hoses and a dual-passage connector at the ECS end and two separate hoses (supply and exhaust) at the PGA end. When not connected to the PGA, the ECS connector end remains attached and the hoses stowed.

Separate water hoses and an electrical cable are connected to the oxygen umbilical by straps secured by snap fasteners. The electrical umbilical carries voice communications and biomedical data, and electrical power for warning-tone impulses. The water hoses circulate water through the LCG.

3-12.8. CREW LIFE SUPPORT.

The crew life support equipment includes food and water, a waste management system, personal hygiene items, and pills for in-flight emergencies. A portable-water unit and food packages contain sufficient life-sustaining supply for completion of the LM mission.

3-12.9. CREW WATER SYSTEM.

The water dispenser assembly consists of a mounting bracket, a coiled hose, and a trigger-actuated water dispenser. The hose and dispenser extend approximately 72 inches to dispense water from the ECS water feed control assembly. The ECS water feed control valve is opened to permit water flow. The dispenser assembly supplies water at +50° to +90° F for drinking or food preparation and fire extinguishing. The water for drinking and food preparation is filtered through a bacteria filter. The water dispenser is inserted directly into the mouth for drinking. Pressing the trigger-type control supplies a thin stream of water for drinking and food preparation. For firefighting, a valve on the dispenser is opened. The valve provides a greater volume of water than that required for drinking and food preparation.

3-12.10. FOOD PREPARATION AND CONSUMPTION.

The astronaut's food supply (approximately 3,500 calories per man per day) includes liquids and solids with adequate nutritional value and low waste content. Food packages are stowed in the LM midsection, on the shelf above PLSS No. 1, the right-hand stowage compartment, and the MESA.

The food is vacuum packed in plastic bags that have one-way poppet valves into which the water dispenser can be inserted. Another valve allows food passage for eating. The food bags are packaged in aluminum-foil-backed plastic bags for stowage and are color coded: red (breakfast), white (lunch), and blue (snacks).

Food preparation involves reconstituting the food with water. The food bag poppet-valve cover is cut with scissors and pushed over the water dispenser nozzle after its protective cover is removed. Pressing the water dispenser trigger releases water. The desired consistency of the food determines the quantity of water added. After withdrawing the water dispenser nozzle, the protective cover is replaced and the dispenser returned to its stowage position. The food bag is kneaded for approximately 3 minutes, after which the food is considered reconstituted. After cutting off the neck of the food bag, food can be squeezed into the mouth through the food-passage valve. A germicide tablet, attached to the outside of the food bag, is inserted into the bag after food consumption, to prevent fermentation and gas formation. The bag is rolled to its smallest size, banded, and placed in the waste disposal compartment.

3-12.11. WASTE MANAGEMENT SYSTEM.

The modified waste management system provides for disposal of body waste through use of a fecal containment system and a urine collection and transfer assembly, and for neutralizing odors. The waste storage container, in the descent stage, accommodates the functions (urine stowage, SLSS and PLSS condensate stowage and gas separation) formerly served by two separate containers. The size and skin thickness (and therefore the weight) of the container is minimized because associated gases and pressure are vented to the lunar environment. Biological contamination of the lunar surface is avoided by incorporation of a bacteria filter at the vent port. An electrical heater prevents ice accumulation, at the inlet and vent ports, which might otherwise result from exposure of the liquids to the lunar vacuum and low temperatures.

A urine receptacle (stowed in the cabin area) collects urine by direct interface with the crew. This assembly consists of a hose and a quick-disconnect, which connects to the PGA waste connector.

The quick-disconnect on the urine receptacle and the overboard dump line contains a self-sealing valve. This feature enables cabin pressure to be retained when the quick-disconnect is broken. In addition, a safety pressure cap on the hose quick-disconnect provides a redundant method of preventing loss of cabin atmosphere.

The hose assembly mates with the vent and drain ports of the PLSS and SLSS for water recharge. PLSS and SLSS condensates are transferred to the descent stage container, using the pressure available at the PLSS and SLSS to effect the transfer.

A folding, stowable seat together with replaceable plastic bags serves as the fecal collection unit. The bags incorporate a sealing device and contain germicide packets and disposable towels.

3-12.12. PERSONAL HYGIENE ITEMS.

Personal hygiene items consist of tissues, towels, and wet facial wipes, chemically treated and sealed in plastic covers. The wipes measure 4 by 4 inches and are folded into 2 inch squares.

3-12.13. MEDICAL EQUIPMENT.

The medical equipment consists of biomedical sensors, personal radiation dosimeters, and emergency medical equipment.

Biomedical sensors gather physiological data for telemetry. Impedance pneumographs continuously record heart beat (EKG) and respiration rate. Each assembly (one for each astronaut) has four electrodes, which contain electrolyte paste; they are attached with tape to the astronaut's body.

Six personal radiation dosimeters are provided for each astronaut. They contain thermoluminescent powder, nuclear emulsions, and film that is sensitive to beta, gamma, and neutron radiation. They are placed on the forehead or right temple, chest, wrist, thigh, and ankle to detect radiation to eyes, bone marrow, and skin. Serious, perhaps critical, damage results if radiation dosage exceeds a predetermined level. For quick, easy reference each astronaut has a dosimeter mounted on his EMU.

The emergency medical equipment consists of a kit of six capsules: four are pain killers (Darvon) and two are pep pills (Dexedrine). The kit is attached to the interior of the flight data file, readily accessible to both astronauts.

3-12.14.   CREW SUPPORT AND RESTRAINT EQUIPMENT.

The crew support and restraint equipment includes armrests, handholds (grips), Velcro
on the floor to interface with the boots, and a restraint assembly operated by a rop-and-
pulley arrangement that secures the astronauts in an upright position under zero-g
conditions.

The armrests, at each astronaut position, provide stability for operation of the thrust/
translation controller assembly and the attitude controller assembly, and restrain the
astronaut laterally.   They are adjustable (four positions) to accommodate the astronaut;
they also have stowed (fully up) and docking (fully down) positions.   The armrests, held
in position by spring-loaded detents, can be moved from the stowed position by grasping
them and applying downward force.   Other positions are selected by pressing latch buttons
on the armrest forward area.   Shock attenuators are built into the armrests for protection
against positive-g forces (lunar landing).   The maximum energy absorption of the armrest
assembly is a 300-pound force, which will cause a 4-inch armrest deflection.

The handholds, at each astronaut station and at various locations around the cabin,
provide support for the upper torso when activity involves turning, reaching, or bending;
they attenuate movement in any direction.   The forward panel handholds are single
upright, peg-type, metal grips.   They are fitted into the forward bulkhead, directly
ahead of the astronauts, and can be grasped with the left or right hand.

The restraint assembly consists of ropes, restraint rings, and a constant-force reel
system.   The ropes attach to D-rings on the PGA sides, waist high.   The constant-force
reel provides a downward force of approximately 30 pounds, it is locked during landing
or docking operations.   When the constant-force reel is locked, the ropes are free to
reel in.   A ratchet stop prevents paying out of the ropes and thus provides zero-g
restraint.   During docking maneuvers, the Commander uses pin adjustments to enable him
to use the crewman optical alignment sight (COAS) at the overhead (docking) window.

3-13.15.   DOCKING AIDS AND TUNNEL HARDWARE.

Docking operations require special equipment and tunnel hardware to effect link up of the
LM with the CSM.   Docking equipment includes the COAS and a docking target.   A drogue
assembly, probe assembly, the CSM forward hatch, and hardware inside the LM tunnel enable
completion of the docking maneuver.

The COAS provides the Commander with gross range cues and closing rate cues during the
docking maneuver.   The closing operation, from 150 feet to contact, is an ocular,
kinesthetic coordination that requires control with minimal use of fuel and time.   The
COAS provides the Commander with a fixed line-of-sight attitude reference image, which
appears to be the same distance away as the target.

The COAS is a collimating instrument.   It weighs approximately 1.5 pounds, is 8 inches
long, and operates from a 28-volt d-c power source.   The COAS consists of a lamp with
an intensity control, a reticle, a barrel-shaped housing and mounting track, and a combiner

and power receptacle. The reticle has vertical and horizontal 10° gradations in a 10° segment of the circular combiner glass, on an elevation scale (right side) of -10° to +31.5°. The COAS is capped and secured to its mount above the left window (position No. 1).

To use the COAS, it is moved from position No. 1 to its mount on the overhead docking window frame (position No. 2) and the panel switch is set from OFF to OVHD. The intensity control is turned clockwise until the reticle appears on the combiner glass; it is adjusted for required brightness.

The docking target permits docking to be accomplished on a three-dimensional alignment basis. The target consists of an inner circle and a standoff cross of black with self-illuminating disks within an outer circumference of white. The target-base diameter is 17.68 inches. The standoff cross is centered 15 inches higher than the base and, as seen at the intercept, is parallel to the X-axis and perpendicular to the Y-axis and the Z-axis.

The drogue assembly consists of a conical structure mounted within the LM docking tunnel. It is secured at three points on the periphery of the tunnel, below the LM docking ring. The LM docking ring is part of the midsection outer structure, concentric with the X-axis. The drogue assembly can be removed from the CSM end or LM end of the tunnel.

Basically, the assembly is a three-section aluminum cone secured with mounting lugs to the LM tunnel ring structure. A lock and release mechanism, on the probe, controls capture of the CSM probe at CSM LM contact. Handles are provided to release the drogue from its tunnel mounts.

The tunnel contains hardware essential to final docking operations. This includes connectors for the electrical umbilicals, docking latches, probe-mounting lugs, tunnel lights, and deadfacing switches.

The probe assembly provides initial CSM-LM coupling and attenuates impact energy imposed by vehicle contact. The probe assembly may be folded for removal and for stowage within either end of the CSM transfer tunnel.

3-12.16. CREW MISCELLANEOUS EQUIPMENT.

Miscellaneous equipment required for completion of crew operations consists of in-flight data with checklists, emergency tool B, and window shades.

The in-flight data are provided in a container in the left-hand midsection. The Commander's checklist is stowed at his station. The in-flight data kit is stowed in a stowage compartment. The packages include the flight plan, experiments data and checklist, mission log and data book, systems data book, and star charts.

Tool B (emergency wrench) is a modified Allen-head L-wrench. It is 6.25 inches long and has a 4.250-inch drive shaft with a 7/16-inch drive. The wrench can apply a torque of 4,175 inch-pounds; it has a ball-lock device to lock the head of the drive shaft. The wrench is stowed on the right side stowage area inside the cabin. It is a contingency tool for use with the probe and drogue, and for opening the CM hatch from outside.

Window shades are used for the overhead (docking) window and forward windows. The window shade material is Aclar. The surface facing outside the cabin has a highly reflective metallic coating. The shade is secured at the bottom (rolled position). To cover the window, the shade is unrolled, flattened against the frame area and secured with snap fasteners.

3-12.17. MODULARIZED EQUIPMENT STOWAGE ASSEMBLY. (See figure 3-12.4.)

The MESA pallet is located in quad 4 of the descent stage. The pallet is deployed by the extravehicular astronaut when the LM is on the lunar surface. It contains fresh PLSS batteries and LiOH cartridges, a TV camera and cable, still cameras, tools for obtaining lunar geological samples, food, film, and containers in which to store the samples. It also has a folding table on which to place the sample return containers. Pallets are provided and are used to transfer the PLSS batteries and the cartridges to the cabin.

The PLSS LiOH cartridges and PLSS batteries are temperature-sensitive items. Their temperature range is +30° to +120° F. To prevent exceeding the minimum allowable temperature, there are heaters near critical items. Low-emissivity coatings on exposed MESA surfaces, and a segmented insulation blanket, are also provided. The temperature in the PLSS cartridge area and in the area that contains the PLSS battery, 70-mm magazine, and stereo camera is telemetered.

3-12-18. APOLLO LUNAR SURFACE EXPERIMENT PACKAGE. (See figure 3-12.5.)

The Apollo Lunar Surface Experiment Package (ALSEP) consists of two packages of scientific instruments and supporting subsystems capable of transmitting scientific data to earth for one year. These data will be used to derive information regarding the composition and structure of the lunar body, its magnetic field, atmosphere and solar wind. Two packages are stowed in quad 2 of the descent stage. The packages are deployed on the lunar surface by the extravehicular astronaut.

ALSEP power is supplied by a radioisotope thermoelectric generator (RTG). Electrical energy is developed through thermoelectric action. The RTG provides a minimum of 16 volts at 56.2 watts to a power-conditioning unit. The radioisotopes fuel capsule emits nuclear radiation and approximately 1,500 thermal watts continuously. The surface temperature of the fuel capsule is approximately 1,400° F. The capsule is stowed in a graphite cask, which is externally mounted on the descent stage. The capsule is removed from the cask and installed in the RTG.

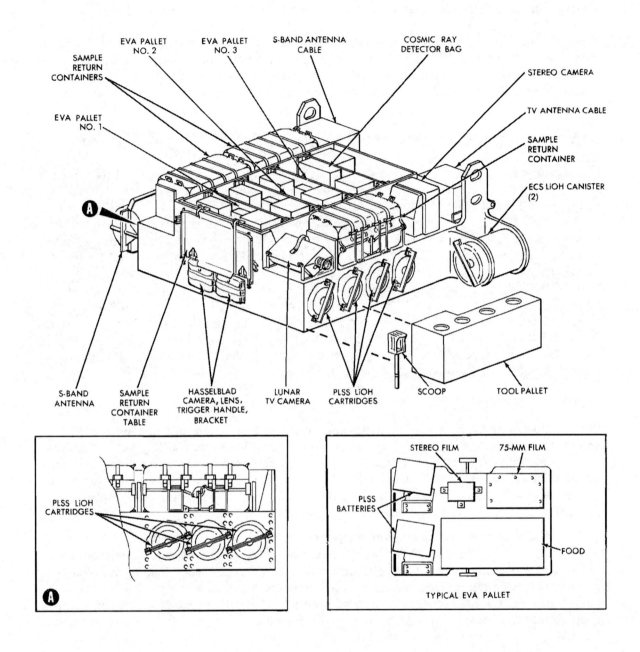

SAMPLE RETURN CONTAINERS

EVA PALLET NO. 2

EVA PALLET NO. 3

S-BAND ANTENNA CABLE

COSMIC RAY DETECTOR BAG

STEREO CAMERA

EVA PALLET NO. 1

TV ANTENNA CABLE

SAMPLE RETURN CONTAINER

ECS LiOH CANISTER (2)

A

S-BAND ANTENNA

SAMPLE RETURN CONTAINER TABLE

HASSELBLAD CAMERA, LENS, TRIGGER HANDLE, BRACKET

LUNAR TV CAMERA

PLSS LiOH CARTRIDGES

SCOOP

TOOL PALLET

PLSS LiOH CARTRIDGES

A

STEREO FILM

75-MM FILM

PLSS BATTERIES

FOOD

TYPICAL EVA PALLET

Figure 3-12.4. Modularized Equipment Stowage Assembly

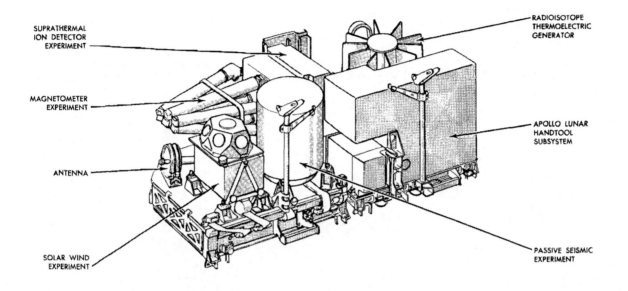

SUPRATHERMAL
ION DETECTOR
EXPERIMENT

MAGNETOMETER
EXPERIMENT

ANTENNA

SOLAR WIND
EXPERIMENT

RADIOISOTOPE
THERMOELECTRIC
GENERATOR

APOLLO LUNAR
HANDTOOL
SUBSYSTEM

PASSIVE SEISMIC
EXPERIMENT

Figure 3-12.5. Apollo Lunar Surface
Experiment Package

3-13. <u>CONTROLS AND DISPLAYS.</u> (See figure 3-13.1.)

The controls and displays enable astronauts to monitor and manage the LM subsystems and to control the vehicle manually during separation, docking, and landing.

In general, the controls and displays are in subsystems groupings located in accordance
with astronaut responsibilities. Certain controls and displays are duplicated to satisfy
mission and/or safety requirements; a system of interlocks prevents simultaneous operation
of these controls. Controls and displays that enable either astronaut to control the vehicle
are centrally located; these are accessible from both flight stations. Controls that could
be operated inadvertently are appropriately guarded.

Annunciator displays go on if malfunctions occur in the vehicle subsystems; at the same
time, two flashing master alarm lights and an alarm tone (in the astronaut headsets) are
activated. Digital and analog displays provide the astronauts with subsystem-status
information such as gas and liquid quantities, pressures, temperatures, and voltages.

There are 13 control and display panels. (See figure 3-13.2.) The main control and
display panels (1 and 2) are canted and centered between the flight stations. A utility
lighting panel is between panels 1 and 2. Panels 3 and 4 are below these panels, within
convenient reach and scan of both astronauts. Panels 5, 8, and 11, and an ORDEAL
panel, are located for use by the Commander. Panels 6, 12, 14, and 16 are located for
use by the LM Pilot. A thrust/translation controller assembly and an attitude controller
assembly are located at the commander and LM Pilot positions.

3-13.1. PANEL 1.

Panel 1 has controls and displays related to flight control, main propulsion, and engine
thrust control. It also has warning lights.

3-13.1.1. <u>Flight Control.</u> The displays related to flight control consist of indicators for
monitoring total attitude, attitude rate, attitude error, range/range rate, altitude/altitude
rate, instantaneous X-axis acceleration, mission elapsed time, and event time. The
controls permit selection of input to the indicator and source of guidance control.

3-13.1.2. <u>Main Propulsion.</u> The displays related to the Main Propulsion Subsystem
consists of indicators for monitoring propellant quantity, propellant, and helium
temperatures, and propellant and helium pressures. The controls permit selection of
inputs to the indicating devices and shut off of helium flow to helium regulators. Talkbacks
(status flags) display valve position.

3-13.1.3. <u>Engine Thrust Control.</u> The controls related to engine thrust permit switching
from automatic to manual throttle control, selection of Commander or LM Pilot attitude
controller to adjust the descent engine thrust level manually, and selection of jets to be
used in X-axis translation maneuvers. They also provide ascent or descent engine-arming
signals to enable engine firing.

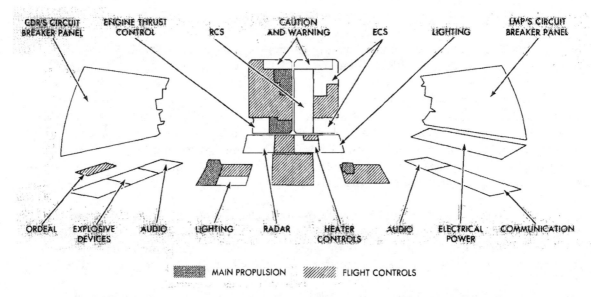

Figure 3-13.1. Controls and Displays - Subsystem Group

3-13.1.4. Warning Lights. The warning lights provide a red indicator to warn of a malfunction that affects astronaut safety and require immediate action to counter the emergency.

3-13.2. PANEL 2.

Panel 2 has controls and displays related to the Reaction Control Subsystem, flight control, and the Environmental Control Subsystem. It also has caution lights.

3-13.2.1. Reaction Control Subsystem. The displays related to the Reaction Control Subsystem (RCS) consists of indicators for monitoring propellant temperature, pressure, and quantity, and critical valve positions. The controls permit selection of inputs to propellant temperature and pressure indicators, and solenoid shut off valves that control propellant flow to the RCS thruster, thruster pair inhibit commands to the LGC, and enable or disable operation of the attitude controller assemblies in the proportional rate mode.

3-13.2.2. Flight Control. The displays related to flight control consists of indicators for monitoring total attitude, attitude rate, attitude error, and instantaneous X-axis acceleration. The controls permit selection of navigational inputs to the indicating devices.

3-13.2.3. Environmental Control Subsystem. The displays related to the Environmental Control Subsystem consists of indicators for monitoring suit and cabin temperature and pressure, partial pressure of carbon dioxide in the atmosphere revitalization section coolant temperature and pressure in the primary coolant loop quantity of oxygen remaining in either descent or either ascent oxygen tank and quantity of water remaining in either descent or either ascent water tank. The O_2/H_2O QTY MON switch enables selection of DES 1, DES 2, ASC 1, or ASC 2. The controls permit selection of either circulating pump in the primary or secondary coolant loop, the suit fan to circulate breathing oxygen in the suit circuit and inputs to the quantity indicators for monitoring.

3-13.2.4. <u>Caution Lights.</u> The caution lights provide a yellow indication to alert the astronauts to a situation or malfunction that is not time-critical to their safety, but requires that they be aware of it.

3-13.3. PANEL 3.

Panel 3 has controls and displays related to radar and heater control, and controls related to stabilization and control, the event timer, and lighting.

3-13.3.1. <u>Radar.</u> The controls related to radar select manual or automatic operation of the rendezvous radar antenna, determine the landing radar antenna position, provide signals to rendezvous and landing radar test circuitry, and provide power to the landing radar.

3-13.3.2. <u>Stabilization and Control.</u> Stabilization and control is part of flight control. The controls related to stabilization and control permit selection of large-amplitude limit cycle for attitude control system, for RCS fuel conservation, or narrow deadband when accurate manual control is required, the rate gyro to be tested in either of three axes, two primary modes (automatic or semiautomatic) of attitude control, and submodes (pulse or divert), individually, in any of the three axes.

3-13.3.3. <u>Heater Control.</u> The controls and displays related to heater control permit monitoring of antennas, RCS quad, and MESA temperatures, and selection of the mode of operation (automatic or manual) of the RCS quad heaters.

3-13.3.4. <u>Event Timer.</u> The controls related to the event timer indicator (panel 1) start and stop the event timer, control the direction in which the event timer counts, and provide slewing function for the event timer.

3-13.3.5. <u>Lighting.</u> The controls related to lighting energize and deenergize the crew compartment floodlights and control their brightness; energize and deenergize the side panel, docking, and tracking lights; and enable lamp and alarm tone test.

3-13.4. PANEL 4.

Panel 4 is related to flight control. It consists of the display and keyboard assembly, which permits the astronauts to load information into the LM guidance computer (LGC), initiate program functions, and perform tests of the LGC and other portions of the Guidance, Navigation, and Control Subsystem. In addition to failures in the LGC, the panel displays indicate program functions being executed by the LGC and specific data selected by the keyboard input. In conjunction with the LGC, the panel supplies indications to the caution and warning lights array.

3-13.5. PANEL 5.

Panel 5 has controls related to the mission timer indicators, main propulsion, and lighting.

3-13.5.1. <u>Mission Timer</u>. The controls related to the mission timer indicator (panel 1) start and stop the mission timer, control the direction in which the mission timer counts, and provides slewing function for the mission timer.

3-13.5.2. <u>Main Propulsion</u>. The controls related to main propulsion permit manual starting and stopping of the descent and ascent engine, and changing the rate of descent, and provide four-jet translation in the X-direction by energizing RCS direct (secondary) coils.

3-13.5.3. <u>Lighting</u>. The controls related to lighting permit control (on and off) of the electroluminescent (EL) lighting for the Commander side panels and adjust the light intensity of all EL panels and numeric displays and of the forward and overhead floodlights.

3-13.6. PANEL 6.

Panel 6 has controls and displays related to flight control and main propulsion.

3-13.6.1. <u>Flight Control</u>. The flight control comprises data entry and display assembly (DEDA). The DEDA is used to control the abort guidance section modes of operation, manually insert data into the abort electronics assembly (AEA), and manually command the contents of a desired AEA memory core to be displayed on the DEDA.

3-13.6.2. <u>Main Propulsion</u>. The control related to main propulsion provides descrete stop signals to the descent and ascent engine.

3-13.7. PANEL 8.

Panel 8 has controls and displays related to descent propulsion, heater controls, explosive devices, and audio.

3-13.7.1. <u>Descent Propulsion</u>. The controls and displays related to descent propulsion permit manual venting of the descent propellant tanks after lunar landing and indicate valve position.

3-13.7.2. <u>Heater Control</u>. The controls related to heater control provide heat to the urine collector (part of the waste management system) to assure proper operation during lunar stay and heat to the MESA during translunar operations and lunar stay.

3-13.7.3. <u>Explosive Devices</u>. The controls related to explosive devices permit arming and disarming of explosive device circuits, deployment of the landing gear, helium pressurization of the ascent and descent propulsion sections and the Reaction Control Subsystem, and staging.

3-13.7.4. <u>Audio</u>. The controls related to audio enable the audio center to receive S-band and VHF/AM voice transmission and route microphone amplifier outputs for transmission via S-band and VHF/AM equipment. The controls also enable reception and transmission of voice via the intercom system, providing a voice conference capability between the extravehicular astronaut and the astronaut in the vehicle.

3-13.8.   PANEL 11.

Panel 11 has five rows of circuit breakers. The circuit breakers are related to the Environmental Control Subsystem; Reaction Control Subsystem; Guidance, Navigation, and Control Subsystem; Main Propulsion Subsystem; Communications Subsystem; Electrical Power Subsystem; Instrumentation Subsystem; and Explosive Devices Subsystem. One circuit breaker, URINE LINE, was added for the modified LM.

3-13.9.   PANEL 12.

Panel 12 has controls related to audio, and controls and displays related to communications and the communications antennas.

3-13.9.1.   Audio.   The controls related to audio enable the audio center to receive S-band and VHF/AM voice transmission and route microphone amplifier outputs for transmission via S-band and VHF/AM equipment. The controls also enable reception and transmission of voice via the intercom system, providing a voice conference capability between the extravehicular astronaut and the astronaut in the vehicle.

3-13.9.2.   Communications.   The controls related to communications enable the Commander and LM Pilot to operate S-band, VHF A, VHF B, telemetry control, the tape recorder, and backup (secondary) S-band equipment. The VHF controls select simplex or duplex voice operation; a squelch control establishes the degree of noise limiting in the operating duplex receiver. The telemetry controls permit transmission of high- or low-bit-rate pulse-code-modulation of biomedical data from either astronaut.

3-13.9.3.   Communications Antennas.   The controls related to the communications antennas permit pointing the S-band steerable antenna at earth and selection of manual track mode, high or low slew rate, and either of two in-flight omnidirectional antennas. The displays indicate azimuth, elevation, and S-band signal strength.

3-13.10.   PANEL 14.

Panel 14 has controls and displays that are related to electrical power. They permit individual selection of outputs from descent batteries No. 1, 2, 3, 4, and lunar battery; ascent batteries No. 5 and 6; enable battery voltage and current monitoring; and indicate the occurrence of battery faults.

3-13.11.   PANEL 16.

Panel 16 has four rows of circuit breakers. The circuit breakers are related to the Environmental Control Subsystem; Reaction Control Subsystem; Guidance, Navigation, and Control Subsystem; Main Propulsion Subsystem; Communications Subsystem; Electrical Power Subsystem; Instrumentation Subsystem; and Explosive Devices Subsystem. One circuit breaker, MESA, was added for the modified LM.

3-13.12.   UTILITY LIGHTS PANEL.

The utility lights panel has switches that control outlets that accept portable utility lights for use in the cabin.

3-13.13.   ORDEAL PANEL.

The ORDEAL panel has controls that permit selection of an alternative attitude for display (in pitch only) on the FDAI's (panels 1 and 2), insertion of the appropriate scale factor into ORDEAL electronics, insertion and maintenance of proper angular offset in FDAI's, and brightness control for ORDEAL EL panel lighting.

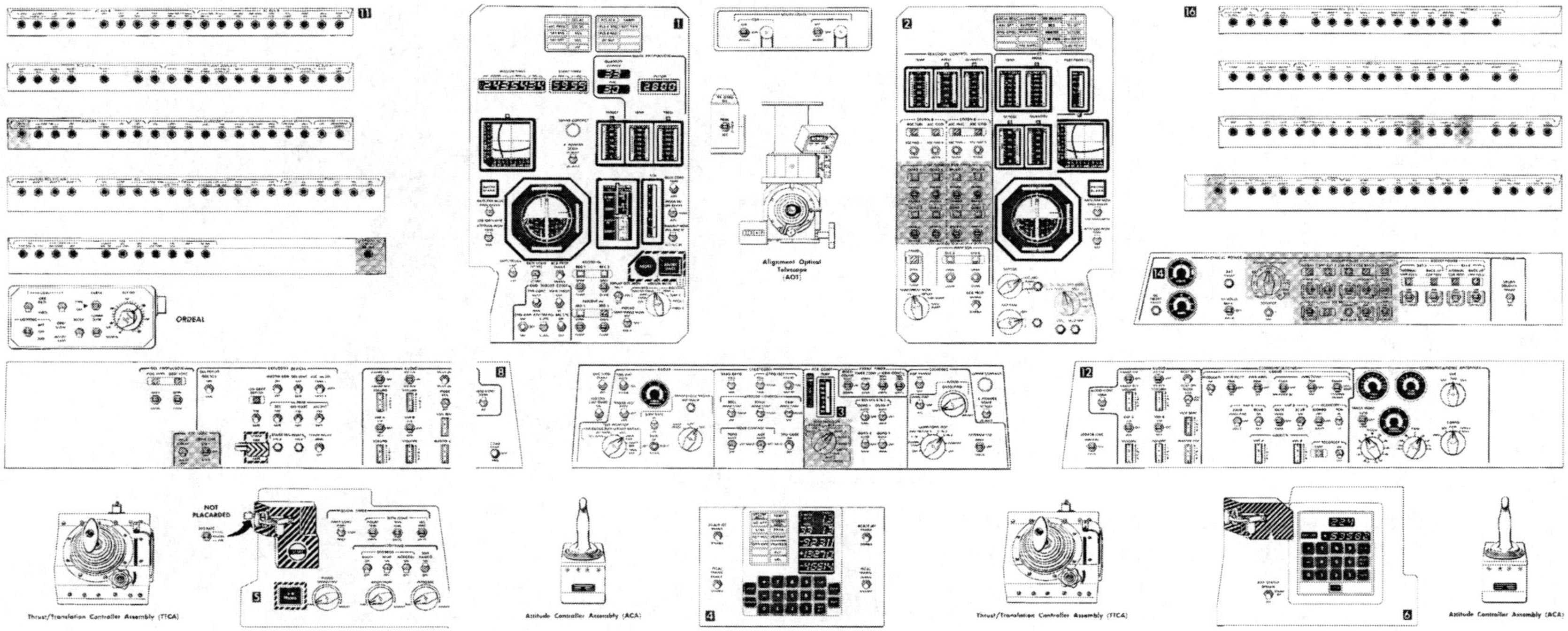
ORDEAL
Alignment Optical Telescope (AOT)
NOT PLACARDED
Thrust/Translation Controller Assembly (TTCA)
Attitude Controller Assembly (ACA)
Thrust/Translation Controller Assembly (TTCA)
Attitude Controller Assembly (ACA)

Figure 3-13.2. Controls and Displays

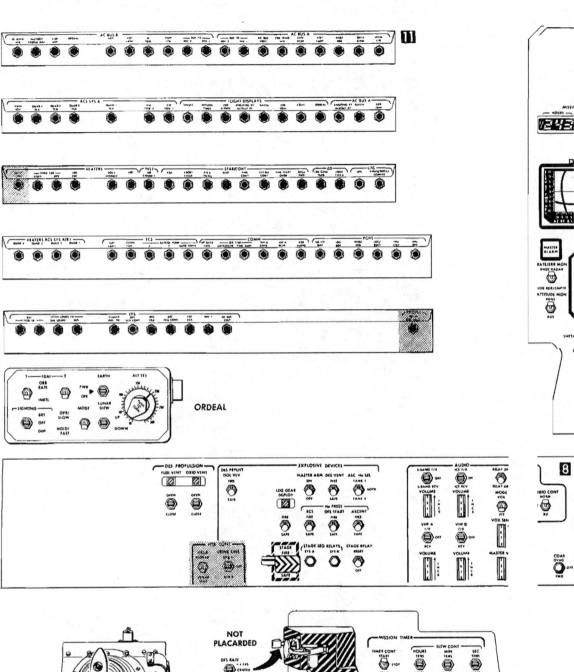

ORDEAL

NOT PLACARDED

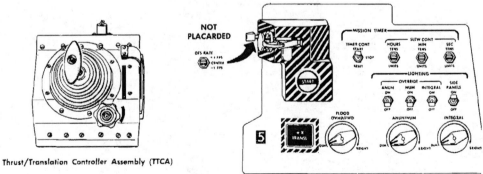

Thrust/Translation Controller Assembly (TTCA)

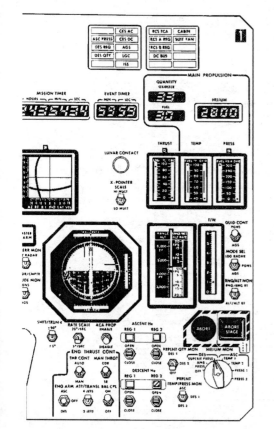

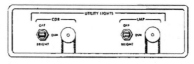

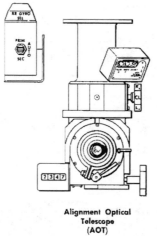

Alignment Optical
Telescope
(AOT)

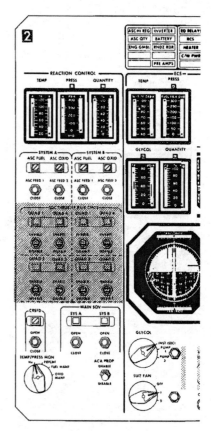

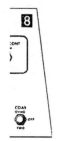

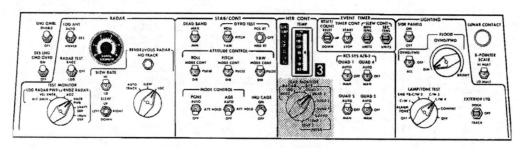

Attitude Controller Assembly (ACA)

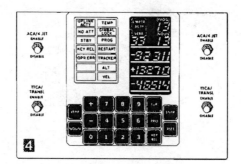

Thrust/Translation Contro

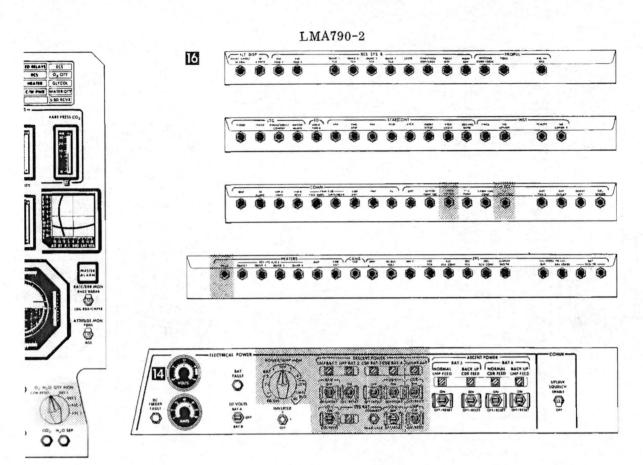

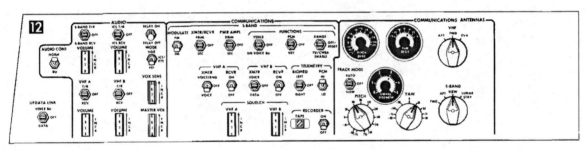

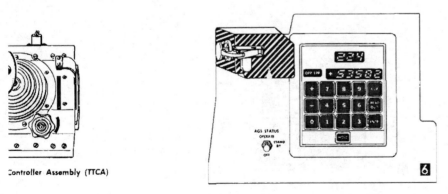

Controller Assembly (TTCA)

Attitude Controller Assembly (ACA)

Figure 3-13.2.   Controls and Displays

SECTION IV

GROUND SUPPORT EQUIPMENT

4-1. <u>GENERAL.</u>

Table 4-1 lists the new ground support equipment (GSE) and other LM GSE end items that will be modified for the LM program.

Table 4-1. New and Modified Ground Support Equipment

Number	Nomenclature	New	Mod
410-1210	Breakout Boxes		X
410-6040	RCS Control Station		X
410-6230	Grumman Cold Flow Facility Wiring		X
410-7900	Installation Stimuli Generator Test Station		X
410-8171	Panel Test Chassis		X
410-8181	Panel Test Chassis		X
410-8330	Auxiliary Switch Relay Box Test Simulator		X
410-9100	Automatic Circuit Analysis Panel Test Chassis Adaptor		X
410-11090	Production Test Cable Set		X
410-11130	Launch Complex 39 Mobile Service Structure Cable Set		X
410-11210	Operation and Checkout Building Cable Set		X
410-11290	Auxiliary Equipment Cable Checkout		X
410-11390	Ascent/Descent Stage Cable Set		X
410-42100	Display and Control Maintenance Test Station		X
410-64711	RCS Valve Position Test Unit		X

Table 4-1. New and Modified Ground Support Equipment (cont)

Number	Nomenclature	New	Mod
410-80051	Electrical Control Assembly Connection Plate		X
410-81131	EPS/GPS Power Interface Checkout Unit		X
410-82200	Power Distribution Maintenance Test Station		X
410-84010	EPS Power Limit and Distribution Assembly		X
420-1250	Transportation Kit		X
420-2060	Cold Flow Workstand Set		X
420-3250	Polarity Test Equipment		X
420-10110	Interior Component Covers Kit		X
420-11004	Guard Panel Set		X
420-11006	Umbilical Lines Guillotine Fixture		X
420-11262	Protective Floor Covers		X
420-11660	LM Environmental Calibration Workstand		X
420-13390	Integrated Workstand		X
420-13500	Descent Stage Transporter		X
420-13650	Descent Stage Workstand		X
420-14102	LM/SLA Servicing Kit		X
420-41510	MESA Handling Equipment	X	
420-52002	Water Umbilical Pull Test Fixture	X	
420-63120	Ascent Stage Engine Plug and Support Kit		X
420-80070	Battery Handling Fixture Outrigger	X	
420-83261	Electrical Circuit Interrupter Support Fixture		X
420-83270	Ascent Stage Battery Installation Fixture		X

Table 4-1. New and Modified Ground Support Equipment (cont)

Number	Nomenclature	New	Mod
420-83281	Battery Stud Simulator		X
420-93176	SLA Upper Work Platform		X
420-93213	SLA Lower Work Platform		X
430-6770	Cold Flow Facility Fluid Distribution System		X
430-52120	Gaseous Oxygen Component Test Stand		X
430-52130	Gaseous Oxygen High Pressure Test Stand		X
430-52200	ECS Checkout Adapter Kit		X
430-52300	Operation and Checkout Building Hose Set		X
430-54200	Gaseous Oxygen Transfer Unit		X
430-54750	Auxiliary Gaseous Oxygen Service Unit		X
430-62230	Porpulsion System Checkout Hose Kit		X
430-62390	Component Test Adapter Set		X
450-9370	Cold Flow Facility Intercabling		X

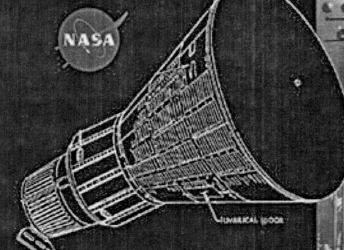

PROJECT MERCURY

FAMILIARIZATION MANUAL

Manned Satellite Capsule

Periscope Film LLC

NASA
PROJECT
GEMINI

FAMILIARIZATION
MANUAL
Manned Satellite Capsule

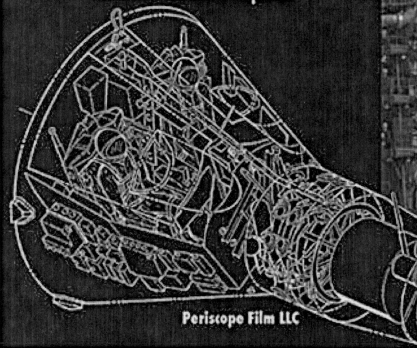

Periscope Film LLC

LMA 790-1

PROJECT APOLLO

lem
LUNAR EXCURSION MODULE

NOW AVAILABLE!

FIRST MANNED LUNAR LANDING
FAMILIARIZATION MANUAL

GRUMMAN AIRCRAFT ENGINEERING CORPORATION • BETHPAGE, L. I., N. Y.

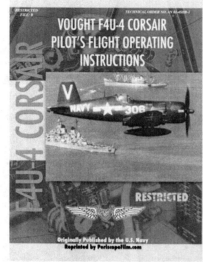

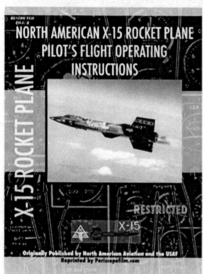
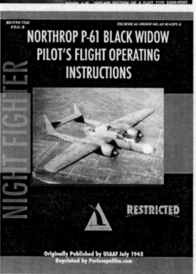

MMS SUBCOURSE NUMBER 151

EDITION CODE 3

NIKE MISSILE
and Test Equipment

NIKE HERCULES

US ARMY

DECLASSIFIED

by U.S. Army Missile and Munitions Center and School
Periscope Film LLC

LMA790-2

CPSIA information can be obtained at www.ICGtesting.com
Printed in the USA
LVOW09s1637271013

358804LV00005B/538/P